Thermodynamic Properties of Water to 1,000°C and 10,000 Bars

C. Wayne Burnham, John R. Holloway and Nicholas F. Davis

Department of Geochemistry and Mineralogy,
College of Earth and Mineral Science,
The Pennsylvania State University,
University Park, Pennsylvania

THE
GEOLOGICAL SOCIETY
OF AMERICA

Special Paper
Number 132

Library of Congress Catalog Card Number 73-96715
S.B.N. 8137-2132-6

Published by
THE GEOLOGICAL SOCIETY OF AMERICA, INC.
Colorado Building, P.O. Box 1719
Boulder, Colorado 80302

Printed in the United States of America

*The printing of this volume has been made possible
through the bequest of
Richard Alexander Fullerton Penrose, Jr.,
and is partially supported by a grant from
The National Science Foundation.*

Acknowledgments

These tables of thermodynamic properties would not have been feasible without our earlier work on the specific volume of water at high temperatures and pressures which was generously supported by the Office of Saline Water, U. S. Department of the Interior, Grant No. 14-01-0001-408. Funds of this grant also were used to defray part of the costs of preparing these tables.

We are indebted to Professor H. L. Barnes, Professor H. P. Taylor, Jr., and Dr. D. H. Lindsley for critically reading the text manuscript and for suggesting ways in which the tables could be made more utilitarian.

Contents

Abstract

The thermodynamic functions for water that are tabulated here consist of: (1) specific volume, (2) Gibbs free energy, (3) entropy, (4) enthalpy, (5) fugacity and (6) fugacity coefficient. They cover the temperature range 20° to 1,000°C in 20° intervals, and the pressure range 100 to 10,000 bars, in 100-bar intervals. In addition, separate tables are presented for the Gibbs free energy at 0.01 and 1.0 bars, as well as for the coefficients in the three empirical equations of state upon which the tabulated values above 1,000 bars are based.

The tabulations up to 800°C and 1,000 bars were obtained, either directly or by computation, from the Steam Tables 1964 (Bain, 1964); all of those above 1,000 bars are based on the specific volume measurements of Burnham, Holloway and Davis (1969). The uncertainty in the tabulated values of specific volume and enthalpy below 1,000 bars is ± 0.1 percent; above 1,000 bars the uncertainty is ± 0.3 percent in specific volume, Gibbs free energy and entropy, ± 0.6 percent in enthalpy and ± 1.3 percent in fugacity and fugacity coefficient.

*Contribution No. 68-19, College of Earth and Mineral Sciences, The Pennsylvania State University. These tables resulted from research supported by Grant No. 14-01-0001-408 from the Office of Saline Water, U. S. Department of the Interior.

Introduction

Knowledge of the thermodynamic properties of water at high temperatures and pressures recently has become increasingly important in many diverse fields. In the fields of igneous and metamorphic petrology, for example, knowledge of the thermodynamic properties of water and aqueous solutions is essential to a quantitative characterization of phase equilibrium relationships involving hydrous minerals, melts and aqueous solutions at high temperatures and pressures. It is therefore essential to an understanding of many petrogenetic processes. As a major step in the furtherance of this understanding, we (1969) have experimentally determined the specific volume of water in the temperature range 20°-900°C and the pressure range 1,000-8,900 bars. These data, together with the tabulated values of specific volume, entropy and enthalpy up to 1,000 bars and 800°C in the National Engineering Laboratory _Steam Tables 1964_ (Bain, 1964), have been used exclusively to derive the tables of thermodynamic properties presented here.

The exclusive use of the _Steam Tables 1964_ for values up to 800°C and 1,000 bars is justified mainly on the basis that: "They are in complete conformity with the new International Skeleton Tables which were formally ratified by the Sixth International Conference on the Properties of Steam, New York, October 1963." (Bain, 1964, p. 1). The tabulated values in the _Steam Tables 1964_ were obtained by fitting empirical equations of state to volumetric data from several sources, perhaps the most extensive of which was the work of Holser and Kennedy (1958, 1959). Beyond 1,000 bars, the upper pressure limit of the _Steam Tables 1964_, the results of

Burnham, Holloway and Davis (1969) were used exclusively because: (1) they represent the most extensive coverage of experimental volumetric determinations available in the pressure-temperature region of these tables, (2) they are internally consistent, and (3) they are in excellent agreement with the low-temperature, high-pressure results of Bridgman (1935), as well as with the 1,000 bar, high-temperature values of Holser and Kennedy (1958, 1959). Moreover, the nature of our experimental apparatus made it possible to eliminate one of the main sources of experimental uncertainty in previous volume measurements at pressures above 1,000 bars and high temperatures; hence, our results in this region should be more reliable.

Pistorius and Sharp (1960) presented tabulations of entropy and Gibbs free energy for water at selected temperatures and pressures up to 1,000^{o}C and 250,000 bars which, in the pressure range of the present tables, were based on the specific volume measurements of Kennedy and co-workers up to 1,400 bars, extrapolations therefrom, and interpolations between these low pressure data and the very high pressure shock-wave results of Rice and Walsh (1957) (Pistorius and Sharp, 1961). Later, Sharp (1962) refined the calculations and expanded the tables to include specific volume, entropy, enthalpy, Gibbs free energy, internal energy and Helmholtz free energy. These latter tabulations are in closer agreement with the present ones than are those of Pistorius and Sharp (1960), but the agreement is good only for values of Gibbs free energy. The largest deviation of Sharp's free-energy values from those in Table 2 is only about 0.5 percent at 1,000^{o}C and 10,000 bars, whereas the deviations in specific volume, entropy and enthalpy at the same temperature and pressure are about 5.1, 2.9 and 6.0 percent, respectively. It should be emphasized, however, that these deviations - which are larger than at any other temperature or pressure - are within the limits of accuracy stated by Sharp (1962, p. 4) for Gibbs free energy, entropy and enthalpy.

Holser (1954) calculated fugacity coefficients of water in the temperature range 200° to 1,000°C and at pressures up to 2,000 bars, using the specific volume data of Kennedy (1950). A comparison with Table 6 reveals general agreement within ± 2.0 percent, except at the extremes in both temperature and pressure. At 200°C and 100 bars, the value in Table 6 is about 14 percent higher than Holser's, whereas at 1,000°C and 2,000 bars it is only 3.2 percent lower. Anderson (1964) also calculated fugacity coefficients in the temperature range 300° to 1,000°C and at pressures up to 10,000 bars, using the Gibbs free energy tabulations of Pistorius and Sharp (1960). In general, they deviate less than 6.0 percent from the values in Table 6, although the deviation reaches nearly 9.0 percent at 1,000°C and 10,000 bars.

The accuracy of the tabulated values for specific volume and enthalpy up to 1,000 bars and 800°C, as stated in the *Steam Tables 1964*, is well within the limits established for the *International Skeleton Tables of the Thermodynamic Properties of Water Substance, 1963* (Bain, 1964). Generally, they do not deviate more than ± 0.1 percent from the internationally accepted values, even at 1,000 bars. The accuracy above 1,000 bars is based on an estimated maximum error of ± 0.3 percent in our specific volume data (Burnham, Holloway and Davis, 1969). This uncertainty leads to a maximum error of ± 0.3 percent in Gibbs free energies and entropies, ± 0.6 percent in enthalpies, and ± 1.3 percent in fugacities and fugacity coefficients.

Entries in the tables above 8,700 bars, at 20°C, are enclosed in parentheses to indicate that they are extrapolated values for metastable liquid water.

Derivation of the Tables

Table 1: Specific volume

The specific volumes of water in Table 1 up to 1,000 bars were taken directly from the Steam Tables 1964; those at 1,000 bars and above were obtained from the three empirical equations of state derived from the experimental data of Burnham, Holloway and Davis (1969). The experimental data extended to 8,400 bars and 900°C; the tabulated values beyond this pressure and temperature were therefore obtained by extrapolation.

The extrapolation to higher pressures was carried out in the following manner. First, all of the measured specific volumes between 4,000 and 8,000 bars were fitted by a ninth degree polynomial in PV as a function of P and T which was used to calculate specific volumes on orthogonal coordinates every 20°C and 100 bars. Next, a quadratic equation was fitted to the 40 calculated values on each 20° isotherm between 20° and 900°C. The 45 resulting quadratic equations were then used to obtain extrapolated specific volumes, at 100 bar intervals, between 8,000 and 10,500 bars.

The extrapolated specific volumes were combined with all the measured values above 1,000 bars and fitted by a ninth degree polynomial of the form:

$$PV = \sum_{j=0}^{9} \sum_{i=0}^{9-j} a_{ij} T^i P^j \qquad (1a)$$

where P is in kilobars and T is in hundreds of degrees C. This equation, the coefficients of which are presented in Table 1a, fits all of the measured values within ± 0.1 percent (one standard deviation), except those within a small wedge-shaped area at low pressures and high temperatures. Data from the sources described by Burnham, Holloway and Davis (1969) in this area, which lies between 1,000 and 1,900 bars, at 900°C, and narrows to between 1,000 and 1,300 bars at 410°C, where it is truncated, were fitted by an eighth degree polynomial of the form:

$$PV = \sum_{j=0}^{8} \sum_{i=0}^{8-j} a_{ij} T^i P^{-j} \tag{1b}$$

This equation, the coefficients of which are presented in Table 1b, fits all of the specific volumes in this area within 0.08 percent (one standard deviation).

Equations (1a) and (1b) were used to calculate all of the specific volumes in Table 1 between 1,000 and 10,000 bars, up to

Table 1a

Coefficients in Equation (1a)

	$a_{ij} \times 10^6$				
i	j=0	j=1	j=2	j=3	j=4
0	-38106.590	1056466.5	-76454.683	709.56600	10454.935
1	66431.473	-42654.409	82193.439	-46017.503	15484.386
2	-89413.365	8922.2030	1971.9748	-4879.6036	1901.7140
3	92076.993	4380.5877	317.79963	-1411.1684	190.87901
4	-47887.922	-618.91487	1904.3321	-111.53384	-12.650068
5	13798.565	-1323.3046	-149.43777	25.569347	-0.34096098
6	-1988.6789	239.12344	-7.5721292	-0.60823080	
7	150.64451	-12.791592	0.57844948		
8	-6.0211303	0.17501650			
9	0.10682110				
i	j=5	j=6	j=7	j=8	j=9
0	-4488.7845	905.86480	-98.313225	5.5147402	-0.12547995
1	-3080.9442	344.62611	-19.805288	0.45366798	
2	-271.55343	15.518783	-0.29266747		
3	-3.3016835	-0.20239777			
4	0.60544695				

Table 1b

Coefficients in Equation (1b)

i	$a_{ij} \times 10^4$				
	j=0	j=1	j=2	j=3	j=4
0	62333.394	2706294.6	-10740139.0	25322920.0	-48726693.0
1	-185685.59	-1095659.1	2043002.0	1963166.8	-5966382.6
2	105601.04	397248.58	-1085579.0	515106.25	-331201.22
3	-30872.023	-54077.509	176931.90	-22749.747	1436.5051
4	4708.6592	2805.9817	-19688.673	1344.6174	282.71453
5	-356.84092	279.03289	1173.3349	-64.912023	
6	4.5184760	-42.896650	-27.139198		
7	1.0665197	1.5083195			
8	-0.048594179				

i	j=5	j=6	j=7	j=8
0	64561026.0	-54093771.0	25488337.0	-5102045.4
1	6729832.0	-3695326.4	797196.76	
2	149491.28	-23568.834		
3	-1668.7313			

900°C. The 20 values between 500° and 900°C on each isobar of Table 1 were fitted by quadratic equations which were extrapolated to obtain additional specific volumes at 20°C intervals to 1,100°C. These extrapolated values were then fitted by a ninth degree polynomial (equation 1c) of the same form as equation (1a). The coefficients of this equation, which also fits the extrapolated values within ± 0.1 percent, are presented in Table 1c.

Thus, all of the specific volumes above 1,000 bars listed in Table 1 were obtained from the three empirical equations of state. These equations, whose functional forms have no theoretical significance, also were used to calculate the other thermodynamic properties presented in Tables 2 through 6. Attempts were made to

Table 1c

Coefficients in Equation (1c)

$a_{ij} \times 10^6$

i	j=0	j=1	j=2	j=3	j=4
0	11116573.0	50844125.0	12381166.0	-2606093.4	895348.18
1	-33982375.0	-48649891.0	-6840564.8	-23371.271	4733.2024
2	19140967.0	19178008.0	2460633.1	-61679.761	-22957.480
3	-3690776.2	-4225522.2	-403459.78	32031.776	1635.3748
4	-95201.802	530054.65	26557.291	-3961.6038	-76.391833
5	163444.50	-34252.319	26.023539	226.45041	1.1103330
6	-30466.416	640.30086	-86.303367	-4.8835854	
7	2734.2941	39.955898	2.9495967		
8	-124.88859	-1.6955926			
9	2.3299454				

i	j=5	j=6	j=7	j=8	j=9
0	-171929.75	15667.520	-652.46589	3.1266465	0.42088718
1	17238.340	-2113.3414	119.25461	-2.8029765	
2	814.90438	-26.937733	0.82580425		
3	-20.163496	-0.43099865			
4	0.98240142				

fit several of the more conventional theoretical and semi-empirical equations of state to the specific volume data, but none of them fit as well as the empirical polynomial functions used.

Table 2: Gibbs Free Energy

The Gibbs free energy (G) values for water in Table 2, up to 1,000 bars, were computed from the specific enthalpies ($\bar{h}$) and entropies ($\bar{s}$) tabulated in the Steam Tables 1964 (Bain, 1964) and the relationship:

$$G = H - TS. \qquad (2a)$$

Above 1,000 bars, differences in free energy (ΔG) along each 20°

isotherm were calculated from:

$$\Delta G = G_P - G_{1000} = \int_{1000}^{P} \overline{V} dP \qquad (2b)$$

where $\overline{V}$ is the molar volume of water expressed as a function of P, at constant T, obtained from equations (1a), (1b) and (1c). The values of G listed in Table 2 were then obtained from:

$$G = G_{1000} + \Delta G, \qquad (2c)$$

where G_{1000} is the Gibbs free energy at T and 1,000 bars from equation (2a). This procedure refers all values of G in Table 2 to a standard state ($G^o = 0$) at the triple point of water (P^o = 0.0061 bar, $T^o = 0.01^oC = 273.16^oK$).

Inasmuch as the Steam Tables 1964 extend only to 800°C, values of G_{1000} up to 1,000°C were obtained by extrapolation, as follows. First, the entropy data from the Steam Tables 1964 in the range 500°-800°C were fitted, by least squares, with the quadratic equation:

$$S_{1000} = 38.579182 - \frac{9868.3171}{T} + \frac{3890077.7}{T^2} \qquad (2d)$$

where T is in °K. The tabulated values of G_{1000} between 800°C (1073°K) and 1,000°C (1273°K) were then obtained from:

$$\Delta G_{1000} = - \int_{1073}^{T} S_{1000} \, dT \qquad (2e)$$

The values of G in Table 2 may be referred to a standard state ($G' = 0$) at 25°C (298.15°K) and one bar with the aid of the following formula:

$$G' = G - 452 + 1.58T \; (^oK). \qquad (2f)$$

Alternatively, they may be referred to a standard state ($G'' = 0$) when $S'' = 0$ at 0°K and $H'' = 0$ at 273.16°K, one bar, by adding the values of -TS in Table 2a. Moreover, by adding another -452 cal $mole^{-1}$ to the values in Table 2b, the Gibbs free energies in Table 2 can be referred to $H' = 0$ at 298.15°K, one bar; the resulting values can then be compared directly to those of Sharp (1962) for G - HO. Also, for some purposes, such as the calculation of hydration-dehydration equilibria using the thermodynamic data

Table 2a

Values for Conversion of Gibbs Free Energy to

S'' = 0 at 0°K and H'' = 0 at 273.16^{o}K

(calories per mole)

T°C	-TS*	T°C	-TS	T°C	-TS	T°C	-TS	T°C	-TS
20	4435	220	7461	420	10487	620	13513	820	16539
40	4738	240	7764	440	10790	640	13816	840	16842
60	5041	260	8067	460	11093	660	14119	860	17145
80	5343	280	8369	480	11395	680	14421	880	17447
100	5646	300	8672	500	11698	700	14724	900	17750
120	5998	320	8974	520	12000	720	15026	920	18052
140	6251	340	9277	540	12303	740	15329	940	18355
160	6554	360	9580	560	12606	760	15632	960	18658
180	6856	380	9882	580	12908	780	15934	980	18960
200	7159	400	10185	600	13211	800	16237	1000	19263

*The value for the third-law entropy (S = 15.13 cal $mole^{-1}$ $^{o}K^{-1}$ at 273.16^{o}K) was taken from Wagman et al. (1965).

Table 2b

The Gibbs Free Energy of Water at 1.0 Bar

(calories per mole)

T°C	$-G_{1.0}$	T°C	$-G_{1.0}$	T°C	$-G_{1.0}$	T°C	$-G_{1.0}$	T°C	$-G_{1.0}$
20	12	220	4255	420	11376	620	19022	820	27097
40	50	240	4938	440	12121	640	19812	840	27925
60	110	260	5628	460	12869	660	20606	860	28757
80	192	280	6329	480	13619	680	21401	880	29593
100	303	300	7029	500	14380	700	22204	900	30431
120	941	320	7739	520	15140	720	23006	920	31273
140	1587	340	8460	540	15909	740	23817	940	32118
160	2242	360	9179	560	16680	760	24630	960	32967
180	2906	380	9907	580	17455	780	25452	980	33818
200	3574	400	10640	600	18236	800	26272	1000	34672

Table 2c

Conversion Factors to Change Standard State for Gibbs Free Energy

New Standard State	Conversion Factor
$G' = 0$ at 298.15°K, 1 bar (Implies $S' = 0$ and $H' = 0$ at 298.15°K, 1 bar).	1.58T (°K) cal $mole^{-1}$ -452 cal $mole^{-1}$.
$G'' = 0$ when $S'' = 0$ at 0°K and $H'' = 0$ at 273.16°K, 1 bar.	-15.13T (°K) cal $mole^{-1}$ (Table 2a).
$G''' = 0$ when $S'' = 0$ at 0°K and $H' = 0$ at 298.15°K, 1 bar.	-452 cal $mole^{-1}$ -15.13T (°K) cal $mole^{-1}$ (Table 2a).
$G'''' = 0$ at T and 1 bar.	Subtract value given in Table 2b.

compiled by Robie and Waldbaum (1968), it is advantageous to choose a sliding reference state of T and one bar. The conversion to this one-bar reference state of a Gibbs free energy value at a given temperature in Table 2 can be readily accomplished by subtracting from it the corresponding value at one bar listed in Table 2b. These values were obtained in exactly the same fashion as the G_{1000} values described above. For convenience, a summary of the conversion factors to be used in changing the standard state for Gibbs free energy is presented in Table 2c.

Table 3: Entropy

The entropy values in Table 3 up to 1,000 bars were obtained, after conversion of units, from the Steam Tables 1964. Above 1,000 bars, the entropy at P relative to that at 1,000 bars was calculated from:

$$\Delta S = \int_{1000}^{P} \left(\frac{\partial \overline{V}}{\partial T}\right)_P dP \tag{3a}$$

where $\overline{V}$ is the molar volume of water obtained from equations (1a)

(1b), and (1c) after conversion of units. The tabulated values of S were then obtained by addition of ΔS to S_{1000} from the Steam Tables 1964 and the extrapolations therefrom according to equation (2d).

The values of S in Table 3 may be referred to a standard state (S' = 0) at 25°C and one bar by subtracting 1.580 cal mole^{-1} °K^{-1}. They also may be referred to S'' = 0 at 0°K by adding 15.130 cal mole^{-1} °K^{-1} (Wagman et al., 1965).

Table 4: Enthalpy

The enthalpy of water up to 1,000 bars was obtained, after conversion of units, from the Steam Tables 1964. Above 1,000 bars, the tabulated values were computed from:

$$H = G + TS. \tag{4a}$$

The values of H in Table 4 may be referred to a standard state (H' = 0) at 298.15°K (25°C) and one bar by subtracting 452 cal mole^{-1}, or to an arbitrary reference value of the ideal gas at 0°K (H''' = 0) by adding 7688 cal mole^{-1} (Sharp, 1962).

Table 5: Fugacity

All of the fugacities (f) listed in Table 5 were calculated from the relation:

$$\ln f/f^{o} = (G - G^{o})/RT \tag{5a}$$

where f^{o} and G^{o} are the fugacity and Gibbs free energy, respectively, in the standard state. The standard-state pressure (P^{o}) chosen for fugacity calculations is 0.01 bars, in order that $f^{o}/P^{o} \rightarrow 1$ at the lower temperatures in Table 5.

The choice of 0.01 bar as the standard-state pressure made it necessary to evaluate $G^{o} = G_{0.01}$ between 800° and 1,000°C. This was done in exactly the same way as for G_{1000}. For 0.01 bar, however, the analogue of equation (2d) is:

$$S_{0.01} = 64.459577 - \frac{19925.784}{T} + \frac{5069124.8}{T^{2}} \tag{5b}$$

Table 5a

The Gibbs Free Energy of Water at 0.01 Bar

(calories per mole)

T^oC	$-G_{0.01}$	T^oC	$-G_{0.01}$	T^oC	$-G_{0.01}$	T^oC	$-G_{0.01}$	T^oC	$-G_{0.01}$
20	506	220	8759	420	17713	620	27188	820	37092
40	1289	240	9628	440	18640	640	28161	840	38103
60	2087	260	10504	460	19570	660	29135	860	39118
80	2889	280	11383	480	20505	680	30116	880	40136
100	3702	300	12270	500	21449	700	31102	900	41158
120	4525	320	13162	520	22395	720	32092	920	42183
140	5357	340	14063	540	23343	740	33085	940	43211
160	6197	360	14965	560	24300	760	34082	960	44242
180	7046	380	15876	580	25255	780	35077	980	45276
200	7898	400	16792	600	26219	800	36085	1000	46313

The values of $G^o = G_{0.01}$, converted in units from the Steam Tables 1964 up to 800°C and extrapolated in temperature with the aid of equation (5b), are presented in Table 5a.

Table 6: Fugacity Coefficient

The entries in this table were obtained by dividing those in Table 5 by the corresponding pressure at each temperature.

References Cited

Anderson, G. M., 1964, The calculated fugacity of water to 1000°C and 10,000 bars: Geoch. et Cosmoch. Acta, v. 28, p. 713-715.

Bain, R. W., 1964, Steam Tables 1964: Natl. Engineering Lab., Edinburgh, Her Majesty's Stationery Office, 147 p.

Bridgman, P. W., 1935, The pressure-volume-temperature relations of the liquid, and the phase diagram of heavy water: Jour. Chem. Physics, v. 3, p. 597-605.

Burnham, C. W., Holloway, J. R., and Davis, N. F., 1969, The specific volume of water in the range 1000 to 8900 bars, 20° to 900°C: Am. Jour. Sci., Schairer vol. 267-A, p. 70-95.

Holser, W. T., 1954, Fugacity of water at high temperatures and pressures: Jour. Phys. Chem., v. 58, p. 316-317.

____________ and Kennedy, G. C., 1958, Properties of water. Part IV. Pressure-volume-temperature relations of water in the range 100-400°C and 100-1400 bars: Am. Jour. Sci., v. 256, p. 744-753.

________________________, 1959, Properties of water. Part V. Pressure-volume-temperature relations of water in the range 400-1000°C and 100-1400 bars: Am. Jour. Sci., v. 256, p. 71-77.

Kennedy, G. C., 1950, Pressure-volume-temperature relations in water at elevated temperatures and pressures: Amer. Jour. Sci., v. 248, p. 540-564.

Pistorius, C. F. W. T., and Sharp, W. E., 1960, Properties of water. Part VI: Entropy and Gibbs free energy of water in the range 10-1000°C and 1-250,000 bars: Am. Jour. Sci., v. 258, p. 757-768.

Pistorius, C. F. W. T., and Sharp, W. E., 1961, Properties of water. Part VI: Entropy and Gibbs free energy of water in the range 10-1000°C and 1-250,000 bars, a clarification: Am. Jour. Sci., v. 259, p. 397-398.

Rice, M. H., and Walsh, J. M., 1957, Equation of state of water to 250 kilobars: Jour. Chem. Physics, v. 26, p. 824-830.

Robie, R. A., and Waldbaum, D. R., 1968, Thermodynamic properties of minerals and related substances at 298.15°K (25.0°C) and one atmosphere (1.013 bars) pressure and at higher temperatures: U. S. Geol. Survey Bull. 1259, 256 pages.

Sharp, W. E., 1962, The thermodynamic functions for water in the range -10 to 1000°C and 1 to 250,000 bars: Livermore, Calif., Univ. Calif. Lawrence Radiation Lab., UCRL-7118, 51 pages.

Wagman, D. D., Evans, W. H., Halow, I., Parker, V. B., Bailey, S. M., and Schumm, R. H., 1965, Selected values of chemical thermodynamic properties, Part I: U.S. Natl. Bur. Standards Tech. Note 270-1, 124 pages.

TABLE 1
SPECIFIC VOLUME IN CUBIC CENTIMETERS PER GRAM

TEMP	PRESSURE IN BARS 100	200	300	400	500	600	700	800
20	0.9971	0.9927	0.9884	0.9843	0.9803	0.9764	0.9726	0.9689
40	1.0034	0.9991	0.9950	0.9909	0.9870	0.9832	0.9795	0.9759
60	1.0127	1.0083	1.0041	1.0000	0.9961	0.9922	0.9884	0.9848
80	1.0243	1.0198	1.0154	1.0112	1.0070	1.0030	0.9991	0.9953
100	1.0383	1.0334	1.0287	1.0243	1.0199	1.0157	1.0116	1.0076
120	1.0550	1.0496	1.0443	1.0394	1.0347	1.0302	1.0259	1.0216
140	1.0741	1.0681	1.0623	1.0568	1.0516	1.0467	1.0419	1.0372
160	1.0958	1.0890	1.0825	1.0764	1.0705	1.0650	1.0597	1.0545
180	1.1203	1.1126	1.1052	1.0981	1.0914	1.0851	1.0792	1.0735
200	1.1483	1.1391	1.1305	1.1225	1.1149	1.1078	1.1010	1.0944
220	1.1807	1.1698	1.1597	1.1501	1.1411	1.1328	1.1250	1.1176
240	1.2187	1.2051	1.1928	1.1813	1.1706	1.1607	1.1516	1.1432
260	1.2647	1.2466	1.2308	1.2167	1.2039	1.1922	1.1814	1.1714
280	1.3221	1.2970	1.2759	1.2578	1.2419	1.2276	1.2147	1.2029
300	1.3970	1.3600	1.3310	1.3070	1.2870	1.2690	1.2530	1.2390
320	19.2300	1.4430	1.4010	1.3660	1.3400	1.3170	1.2980	1.2800
340	21.4600	1.5700	1.4930	1.4410	1.4060	1.3750	1.3500	1.3280
360	23.2900	1.8220	1.6290	1.5410	1.4860	1.4460	1.4110	1.3840
380	24.9100	8.2560	1.8790	1.6880	1.5940	1.5310	1.4840	1.4470
400	26.3900	9.9500	2.8180	1.9140	1.7330	1.6360	1.5700	1.5200
420	27.7700	11.1900	4.9200	2.3660	1.9420	1.7750	1.6760	1.6070
440	29.0900	12.2300	6.2320	3.2050	2.2690	1.9660	1.8120	1.7140
460	30.3400	13.1500	7.1950	4.1420	2.7450	2.2280	1.9870	1.8440
480	31.5600	13.9900	7.9900	4.9420	3.3130	2.5630	2.2050	2.0030
500	32.7500	14.7700	8.6840	5.6180	3.8820	2.9500	2.4660	2.1910
520	33.9000	15.5000	9.3110	6.2070	4.4090	3.3560	2.7570	2.4040
540	35.0300	16.2000	9.8890	6.7340	4.8880	3.7550	3.0630	2.6370
560	36.1400	16.8700	10.4300	7.2170	5.3260	4.1350	3.3730	2.8820
580	37.2400	17.5200	10.9400	7.6650	5.7300	4.4930	3.6760	3.1300
600	38.3100	18.1500	11.4300	8.0870	6.1080	4.8300	3.9690	3.3780
620	39.3800	18.7700	11.9000	8.4870	6.4630	5.1480	4.2500	3.6210
640	40.4300	19.3700	12.3600	8.8680	6.7990	5.4490	4.5180	3.8560
660	41.4700	19.9500	12.8000	9.2350	7.1200	5.7350	4.7750	4.0850
680	42.5100	20.5300	13.2200	9.5890	7.4280	6.0100	5.0210	4.3050
700	43.5300	21.1000	13.6400	9.9320	7.7250	6.2730	5.2570	4.5180
720	44.5500	21.6600	14.0500	10.2600	8.0110	6.5270	5.4850	4.7230
740	45.5600	22.2100	14.4500	10.5900	8.2890	6.7720	5.7050	4.9220
760	46.5700	22.7600	14.8400	10.9100	8.5590	7.0100	5.9180	5.1150
780	47.5700	23.3000	15.2300	11.2200	8.8220	7.2410	6.1250	5.3020
800	48.5600	23.8300	15.6100	11.5200	9.0790	7.4660	6.3260	5.4840

TABLE 1
SPECIFIC VOLUME IN CUBIC CENTIMETERS PER GRAM

	PRESSURE IN BARS							
TEMP	900	1000	1100	1200	1300	1400	1500	1600
20	0.9653	0.9614	0.9587	0.9559	0.9529	0.9498	0.9468	0.9438
40	0.9724	0.9704	0.9676	0.9646	0.9616	0.9585	0.9554	0.9523
60	0.9812	0.9790	0.9762	0.9731	0.9700	0.9669	0.9638	0.9607
80	0.9917	0.9886	0.9857	0.9826	0.9794	0.9762	0.9731	0.9699
100	1.0037	0.9999	0.9968	0.9936	0.9903	0.9870	0.9837	0.9804
120	1.0174	1.0131	1.0097	1.0063	1.0028	0.9994	0.9959	0.9925
140	1.0327	1.0282	1.0245	1.0208	1.0171	1.0134	1.0097	1.0061
160	1.0495	1.0450	1.0410	1.0370	1.0330	1.0290	1.0250	1.0211
180	1.0679	1.0635	1.0591	1.0547	1.0503	1.0460	1.0417	1.0375
200	1.0881	1.0836	1.0786	1.0737	1.0690	1.0643	1.0596	1.0551
220	1.1106	1.1051	1.0995	1.0942	1.0889	1.0838	1.0788	1.0738
240	1.1353	1.1285	1.1221	1.1160	1.1102	1.1046	1.0991	1.0937
260	1.1622	1.1539	1.1464	1.1395	1.1330	1.1268	1.1207	1.1148
280	1.1921	1.1819	1.1730	1.1650	1.1576	1.1506	1.1438	1.1373
300	1.2250	1.2131	1.2024	1.1929	1.1843	1.1762	1.1686	1.1613
320	1.2660	1.2484	1.2352	1.2238	1.2135	1.2042	1.1954	1.1871
340	1.3100	1.2884	1.2720	1.2581	1.2458	1.2348	1.2246	1.2151
360	1.3600	1.3341	1.3137	1.2966	1.2817	1.2685	1.2565	1.2455
380	1.4160	1.3864	1.3610	1.3398	1.3217	1.3058	1.2916	1.2787
400	1.4810	1.4462	1.4146	1.3885	1.3664	1.3472	1.3303	1.3150
420	1.5550	1.5117	1.4715	1.4432	1.4170	1.3931	1.3729	1.3549
440	1.6430	1.5859	1.5371	1.5021	1.4709	1.4439	1.4197	1.3984
460	1.7480	1.6732	1.6128	1.5688	1.5311	1.4990	1.4712	1.4460
480	1.8710	1.7756	1.7000	1.6444	1.5984	1.5601	1.5273	1.4978
500	2.0160	1.8936	1.7992	1.7295	1.6731	1.6271	1.5884	1.5539
520	2.1800	2.0263	1.9102	1.8239	1.7554	1.7003	1.6545	1.6144
540	2.3610	2.1723	2.0321	1.9273	1.8450	1.7796	1.7259	1.6792
560	2.5560	2.3294	2.1636	2.0389	1.9415	1.8647	1.8021	1.7483
580	2.7580	2.4952	2.3031	2.1575	2.0443	1.9552	1.8830	1.8214
600	2.9640	2.6672	2.4490	2.2822	2.1526	2.0507	1.9683	1.8993
620	3.1710	2.8430	2.5995	2.4117	2.2654	2.1504	2.0575	1.9799
640	3.3760	3.0203	2.7529	2.5447	2.3819	2.2538	2.1502	2.0637
660	3.5770	3.1971	2.9076	2.6799	2.5011	2.3599	2.2456	2.1502
680	3.7730	3.3717	3.0622	2.8162	2.6221	2.4682	2.3433	2.2389
700	3.9640	3.5430	3.2156	2.9527	2.7439	2.5777	2.4424	2.3292
720	4.1500	3.7102	3.3669	3.0884	2.8659	2.6879	2.5425	2.4207
740	4.3300	3.8729	3.5155	3.2227	2.9874	2.7982	2.6431	2.5127
760	4.5050	4.0313	3.6610	3.3553	3.1079	2.9081	2.7436	2.6051
780	4.6750	4.1856	3.8035	3.4857	3.2272	3.0173	2.8439	2.6974
800	4.8410	4.3365	3.9432	3.6142	3.3451	3.1257	2.9436	2.7894
820		4.4849	4.0803	3.7407	3.4617	3.2332	3.0428	2.8812
840		4.6312	4.2153	3.8656	3.5772	3.3399	3.1416	2.9727
860		4.7761	4.3486	3.9892	3.6918	3.4461	3.2400	3.0640
880		4.9195	4.4803	4.1116	3.8057	3.5519	3.3381	3.1551
900		5.0947	4.6150	4.2282	3.9111	3.6473	3.4251	3.2356
920		5.2219	4.7322	4.3369	4.0125	3.7424	3.5145	3.3200
940		5.3423	4.8443	4.4419	4.1112	3.8355	3.6026	3.4034
960		5.4576	4.9528	4.5444	4.2083	3.9276	3.6900	3.4865
980		5.5704	5.0600	4.6463	4.3052	4.0198	3.7778	3.5700
1000		5.6843	5.1687	4.7500	4.4042	4.1141	3.8675	3.6553

TABLE 1
SPECIFIC VOLUME IN CUBIC CENTIMETERS PER GRAM

TEMP	PRESSURE IN BARS 1700	1800	1900	2000	2100	2200	2300	2400
20	0.9408	0.9379	0.9351	0.9323	0.9296	0.9270	0.9245	0.9221
40	0.9493	0.9464	0.9435	0.9407	0.9380	0.9354	0.9329	0.9304
60	0.9577	0.9547	0.9518	0.9490	0.9463	0.9437	0.9411	0.9386
80	0.9668	0.9638	0.9609	0.9580	0.9552	0.9525	0.9499	0.9473
100	0.9772	0.9741	0.9710	0.9680	0.9651	0.9623	0.9596	0.9569
120	0.9891	0.9858	0.9825	0.9794	0.9763	0.9733	0.9704	0.9676
140	1.0025	0.9989	0.9954	0.9921	0.9888	0.9856	0.9825	0.9795
160	1.0172	1.0134	1.0097	1.0060	1.0025	0.9990	0.9956	0.9924
180	1.0333	1.0292	1.0251	1.0212	1.0173	1.0136	1.0099	1.0064
200	1.0505	1.0461	1.0417	1.0374	1.0333	1.0292	1.0252	1.0214
220	1.0689	1.0641	1.0594	1.0547	1.0502	1.0458	1.0415	1.0373
240	1.0884	1.0832	1.0780	1.0730	1.0681	1.0633	1.0586	1.0541
260	1.1090	1.1033	1.0978	1.0923	1.0870	1.0818	1.0767	1.0718
280	1.1309	1.1247	1.1186	1.1127	1.1069	1.1012	1.0957	1.0903
300	1.1542	1.1474	1.1407	1.1342	1.1279	1.1217	1.1157	1.1099
320	1.1792	1.1716	1.1642	1.1570	1.1501	1.1433	1.1368	1.1304
340	1.2061	1.1975	1.1892	1.1813	1.1736	1.1662	1.1591	1.1521
360	1.2351	1.2254	1.2161	1.2072	1.1987	1.1905	1.1826	1.1750
380	1.2667	1.2555	1.2450	1.2350	1.2255	1.2164	1.2076	1.1992
400	1.3011	1.2882	1.2762	1.2649	1.2541	1.2439	1.2342	1.2249
420	1.3386	1.3236	1.3098	1.2969	1.2848	1.2733	1.2625	1.2521
440	1.3793	1.3620	1.3461	1.3314	1.3176	1.3047	1.2925	1.2810
460	1.4237	1.4035	1.3852	1.3684	1.3527	1.3382	1.3245	1.3116
480	1.4717	1.4484	1.4273	1.4080	1.3902	1.3738	1.3584	1.3440
500	1.5236	1.4966	1.4724	1.4503	1.4302	1.4116	1.3943	1.3782
520	1.5793	1.5483	1.5205	1.4954	1.4726	1.4516	1.4323	1.4143
540	1.6389	1.6034	1.5717	1.5433	1.5175	1.4939	1.4722	1.4522
560	1.7022	1.6618	1.6259	1.5938	1.5648	1.5384	1.5142	1.4919
580	1.7692	1.7235	1.6830	1.6469	1.6144	1.5850	1.5581	1.5334
600	1.8396	1.7882	1.7429	1.7026	1.6664	1.6336	1.6039	1.5766
620	1.9132	1.8558	1.8053	1.7605	1.7204	1.6842	1.6514	1.6214
640	1.9896	1.9260	1.8701	1.8206	1.7764	1.7366	1.7006	1.6678
660	2.0686	1.9985	1.9370	1.8826	1.8342	1.7906	1.7513	1.7155
680	2.1498	2.0729	2.0058	1.9464	1.8935	1.8461	1.8034	1.7646
700	2.2325	2.1490	2.0760	2.0116	1.9543	1.9030	1.8567	1.8148
720	2.3164	2.2264	2.1476	2.0781	2.0163	1.9610	1.9112	1.8662
740	2.4011	2.3048	2.2201	2.1455	2.0792	2.0200	1.9667	1.9185
760	2.4863	2.3832	2.2933	2.2137	2.1430	2.0798	2.0230	1.9717
780	2.5715	2.4623	2.3670	2.2824	2.2074	2.1404	2.0801	2.0256
800	2.6568	2.5415	2.4408	2.3515	2.2722	2.2014	2.1377	2.0801
820	2.7419	2.6208	2.5145	2.4206	2.3373	2.2628	2.1958	2.1352
840	2.8269	2.7000	2.5890	2.4898	2.4025	2.3244	2.2542	2.1907
860	2.9119	2.7793	2.6633	2.5587	2.4677	2.3862	2.3128	2.2464
880	2.9967	2.8586	2.7376	2.6274	2.5328	2.4480	2.3716	2.3023
900	3.0724	2.9305	2.8060	2.6959	2.5977	2.5097	2.4303	2.3583
920	3.1522	3.0061	2.8777	2.7640	2.6625	2.5713	2.4890	2.4142
940	3.2314	3.0812	2.9491	2.8319	2.7271	2.6328	2.5475	2.4699
960	3.3103	3.1563	3.0205	2.8997	2.7916	2.6941	2.6057	2.5252
980	3.3898	3.2318	3.0922	2.9678	2.8562	2.7553	2.6637	2.5801
1000	3.4707	3.3086	3.1649	3.0365	2.9211	2.8166	2.7215	2.6344

TABLE 1
SPECIFIC VOLUME IN CUBIC CENTIMETERS PER GRAM

	PRESSURE IN BARS							
TEMP	2500	2600	2700	2800	2900	3000	3100	3200
20	0.9197	0.9173	0.9150	0.9128	0.9106	0.9085	0.9064	0.9043
40	0.9280	0.9257	0.9234	0.9212	0.9190	0.9168	0.9147	0.9127
60	0.9362	0.9338	0.9316	0.9293	0.9271	0.9250	0.9229	0.9208
80	0.9449	0.9425	0.9401	0.9379	0.9356	0.9335	0.9313	0.9292
100	0.9544	0.9519	0.9495	0.9471	0.9448	0.9426	0.9404	0.9382
120	0.9649	0.9623	0.9597	0.9572	0.9548	0.9525	0.9502	0.9479
140	0.9765	0.9737	0.9710	0.9683	0.9657	0.9632	0.9608	0.9584
160	0.9892	0.9862	0.9832	0.9803	0.9775	0.9748	0.9722	0.9697
180	1.0029	0.9996	0.9964	0.9933	0.9902	0.9873	0.9845	0.9817
200	1.0176	1.0140	1.0105	1.0071	1.0038	1.0006	0.9975	0.9945
220	1.0332	1.0293	1.0254	1.0217	1.0181	1.0147	1.0113	1.0080
240	1.0497	1.0454	1.0412	1.0372	1.0333	1.0295	1.0258	1.0223
260	1.0670	1.0623	1.0578	1.0534	1.0491	1.0450	1.0410	1.0372
280	1.0851	1.0801	1.0751	1.0704	1.0658	1.0613	1.0570	1.0528
300	1.1042	1.0987	1.0934	1.0882	1.0832	1.0783	1.0737	1.0691
320	1.1243	1.1183	1.1125	1.1069	1.1014	1.0962	1.0911	1.0862
340	1.1454	1.1389	1.1326	1.1264	1.1206	1.1148	1.1093	1.1040
360	1.1676	1.1605	1.1536	1.1470	1.1406	1.1344	1.1284	1.1227
380	1.1911	1.1833	1.1758	1.1686	1.1616	1.1549	1.1484	1.1421
400	1.2160	1.2074	1.1992	1.1913	1.1837	1.1763	1.1693	1.1625
420	1.2423	1.2328	1.2238	1.2151	1.2068	1.1988	1.1912	1.1838
440	1.2701	1.2596	1.2497	1.2402	1.2311	1.2224	1.2140	1.2060
460	1.2994	1.2879	1.2769	1.2665	1.2565	1.2470	1.2379	1.2292
480	1.3305	1.3177	1.3056	1.2941	1.2832	1.2728	1.2629	1.2534
500	1.3632	1.3490	1.3356	1.3230	1.3110	1.2997	1.2888	1.2785
520	1.3976	1.3819	1.3671	1.3532	1.3401	1.3277	1.3159	1.3047
540	1.4336	1.4163	1.4000	1.3847	1.3704	1.3568	1.3439	1.3318
560	1.4713	1.4522	1.4343	1.4175	1.4018	1.3870	1.3730	1.3598
580	1.5107	1.4896	1.4699	1.4516	1.4344	1.4183	1.4031	1.3888
600	1.5516	1.5284	1.5069	1.4869	1.4682	1.4506	1.4341	1.4186
620	1.5939	1.5686	1.5451	1.5233	1.5030	1.4840	1.4661	1.4494
640	1.6377	1.6101	1.5846	1.5609	1.5389	1.5183	1.4991	1.4810
660	1.6828	1.6528	1.6251	1.5995	1.5757	1.5536	1.5329	1.5135
680	1.7292	1.6967	1.6668	1.6392	1.6136	1.5898	1.5675	1.5468
700	1.7766	1.7417	1.7095	1.6799	1.6524	1.6269	1.6031	1.5808
720	1.8252	1.7877	1.7532	1.7215	1.6921	1.6648	1.6394	1.6157
740	1.8747	1.8346	1.7978	1.7639	1.7326	1.7035	1.6765	1.6513
760	1.9250	1.8823	1.8432	1.8072	1.7739	1.7431	1.7144	1.6876
780	1.9761	1.9309	1.8894	1.8512	1.8160	1.7833	1.7530	1.7247
800	2.0278	1.9801	1.9363	1.8960	1.8587	1.8242	1.7922	1.7623
820	2.0801	2.0299	1.9837	1.9413	1.9021	1.8658	1.8320	1.8006
840	2.1329	2.0801	2.0317	1.9871	1.9459	1.9078	1.8723	1.8392
860	2.1860	2.1307	2.0800	2.0333	1.9901	1.9501	1.9129	1.8782
880	2.2393	2.1816	2.1286	2.0797	2.0346	1.9926	1.9537	1.9173
900	2.2926	2.2325	2.1772	2.1262	2.0790	2.0351	1.9943	1.9563
920	2.3459	2.2833	2.2257	2.1724	2.1231	2.0773	2.0347	1.9949
940	2.3989	2.3337	2.2737	2.2182	2.1667	2.1189	2.0743	2.0327
960	2.4514	2.3836	2.3211	2.2632	2.2094	2.1594	2.1128	2.0692
980	2.5034	2.4327	2.3675	2.3070	2.2508	2.1985	2.1497	2.1041
1000	2.5545	2.4808	2.4126	2.3493	2.2905	2.2357	2.1845	2.1367

TABLE 1
SPECIFIC VOLUME IN CUBIC CENTIMETERS PER GRAM

TEMP	PRESSURE IN BARS 3300	3400	3500	3600	3700	3800	3900	4000
20	0.9023	0.9002	0.8982	0.8963	0.8943	0.8924	0.8904	0.8886
40	0.9106	0.9086	0.9066	0.9046	0.9027	0.9008	0.8988	0.8969
60	0.9188	0.9168	0.9148	0.9128	0.9109	0.9090	0.9070	0.9051
80	0.9272	0.9252	0.9232	0.9212	0.9192	0.9173	0.9154	0.9135
100	0.9361	0.9340	0.9320	0.9300	0.9280	0.9261	0.9241	0.9222
120	0.9457	0.9436	0.9415	0.9394	0.9373	0.9353	0.9333	0.9313
140	0.9561	0.9538	0.9516	0.9494	0.9473	0.9451	0.9431	0.9410
160	0.9672	0.9648	0.9624	0.9601	0.9578	0.9556	0.9534	0.9512
180	0.9790	0.9764	0.9739	0.9714	0.9690	0.9666	0.9643	0.9620
200	0.9916	0.9888	0.9860	0.9834	0.9807	0.9782	0.9757	0.9733
220	1.0049	1.0018	0.9988	0.9959	0.9931	0.9904	0.9877	0.9851
240	1.0188	1.0155	1.0123	1.0091	1.0061	1.0031	1.0003	0.9974
260	1.0335	1.0298	1.0263	1.0230	1.0197	1.0165	1.0134	1.0103
280	1.0488	1.0448	1.0410	1.0374	1.0338	1.0304	1.0270	1.0237
300	1.0647	1.0605	1.0564	1.0524	1.0486	1.0448	1.0412	1.0377
320	1.0814	1.0768	1.0724	1.0681	1.0640	1.0599	1.0560	1.0523
340	1.0989	1.0939	1.0891	1.0845	1.0800	1.0756	1.0714	1.0674
360	1.1171	1.1117	1.1065	1.1015	1.0967	1.0920	1.0875	1.0831
380	1.1361	1.1303	1.1247	1.1193	1.1141	1.1090	1.1042	1.0994
400	1.1560	1.1497	1.1436	1.1378	1.1322	1.1267	1.1215	1.1164
420	1.1767	1.1699	1.1634	1.1571	1.1510	1.1451	1.1395	1.1340
440	1.1983	1.1910	1.1839	1.1771	1.1705	1.1642	1.1582	1.1523
460	1.2209	1.2129	1.2052	1.1979	1.1908	1.1840	1.1775	1.1712
480	1.2443	1.2357	1.2274	1.2195	1.2119	1.2045	1.1975	1.1908
500	1.2687	1.2593	1.2504	1.2418	1.2336	1.2257	1.2182	1.2110
520	1.2940	1.2839	1.2742	1.2649	1.2561	1.2476	1.2395	1.2318
540	1.3202	1.3092	1.2988	1.2888	1.2793	1.2702	1.2615	1.2532
560	1.3473	1.3354	1.3241	1.3134	1.3031	1.2934	1.2841	1.2751
580	1.3752	1.3624	1.3502	1.3387	1.3277	1.3172	1.3072	1.2977
600	1.4040	1.3902	1.3771	1.3647	1.3529	1.3417	1.3310	1.3208
620	1.4336	1.4187	1.4047	1.3914	1.3787	1.3667	1.3553	1.3445
640	1.4641	1.4481	1.4330	1.4187	1.4052	1.3924	1.3802	1.3687
660	1.4953	1.4781	1.4620	1.4468	1.4323	1.4187	1.4057	1.3934
680	1.5273	1.5089	1.4917	1.4754	1.4601	1.4455	1.4318	1.4187
700	1.5600	1.5405	1.5221	1.5048	1.4885	1.4730	1.4584	1.4445
720	1.5935	1.5727	1.5532	1.5348	1.5174	1.5011	1.4856	1.4709
740	1.6277	1.6057	1.5849	1.5654	1.5470	1.5297	1.5133	1.4978
760	1.6627	1.6393	1.6173	1.5967	1.5773	1.5589	1.5416	1.5252
780	1.6983	1.6735	1.6503	1.6285	1.6080	1.5887	1.5704	1.5531
800	1.7345	1.7084	1.6839	1.6609	1.6393	1.6189	1.5997	1.5815
820	1.7712	1.7437	1.7180	1.6938	1.6710	1.6496	1.6293	1.6102
840	1.8084	1.7795	1.7524	1.7270	1.7031	1.6805	1.6593	1.6392
860	1.8458	1.8155	1.7871	1.7604	1.7353	1.7116	1.6893	1.6683
880	1.8834	1.8516	1.8218	1.7938	1.7675	1.7427	1.7193	1.6972
900	1.9207	1.8875	1.8562	1.8269	1.7993	1.7733	1.7489	1.7257
920	1.9577	1.9228	1.8901	1.8594	1.8305	1.8033	1.7777	1.7535
940	1.9937	1.9572	1.9230	1.8909	1.8607	1.8322	1.8054	1.7801
960	2.0285	1.9903	1.9545	1.9208	1.8892	1.8595	1.8315	1.8051
980	2.0614	2.0214	1.9839	1.9487	1.9156	1.8845	1.8553	1.8277
1000	2.0919	2.0500	2.0107	1.9738	1.9392	1.9067	1.8761	1.8474

TABLE 1
SPECIFIC VOLUME IN CUBIC CENTIMETERS PER GRAM

	PRESSURE IN BARS							
TEMP	4100	4200	4300	4400	4500	4600	4700	4800
20	0.8867	0.8848	0.8830	0.8811	0.8793	0.8776	0.8758	0.8741
40	0.8951	0.8932	0.8913	0.8895	0.8877	0.8859	0.8841	0.8823
60	0.9033	0.9014	0.8995	0.8977	0.8959	0.8941	0.8923	0.8905
80	0.9116	0.9097	0.9079	0.9060	0.9042	0.9024	0.9006	0.8988
100	0.9203	0.9184	0.9165	0.9147	0.9128	0.9110	0.9091	0.9073
120	0.9294	0.9275	0.9255	0.9236	0.9218	0.9199	0.9180	0.9162
140	0.9390	0.9370	0.9350	0.9331	0.9311	0.9292	0.9273	0.9254
160	0.9491	0.9470	0.9450	0.9429	0.9409	0.9389	0.9369	0.9350
180	0.9597	0.9575	0.9554	0.9532	0.9511	0.9490	0.9470	0.9449
200	0.9709	0.9685	0.9662	0.9640	0.9617	0.9595	0.9574	0.9552
220	0.9825	0.9800	0.9776	0.9752	0.9728	0.9705	0.9682	0.9660
240	0.9947	0.9920	0.9894	0.9868	0.9843	0.9819	0.9794	0.9770
260	1.0074	1.0045	1.0017	0.9990	0.9963	0.9936	0.9911	0.9885
280	1.0206	1.0175	1.0145	1.0115	1.0087	1.0059	1.0031	1.0004
300	1.0343	1.0310	1.0278	1.0246	1.0215	1.0185	1.0156	1.0127
320	1.0486	1.0450	1.0416	1.0382	1.0349	1.0317	1.0286	1.0255
340	1.0634	1.0596	1.0559	1.0523	1.0488	1.0453	1.0420	1.0387
360	1.0789	1.0747	1.0708	1.0669	1.0631	1.0595	1.0559	1.0524
380	1.0949	1.0905	1.0862	1.0820	1.0780	1.0741	1.0703	1.0666
400	1.1115	1.1068	1.1022	1.0978	1.0935	1.0893	1.0852	1.0813
420	1.1288	1.1237	1.1188	1.1140	1.1094	1.1050	1.1006	1.0964
440	1.1467	1.1412	1.1360	1.1309	1.1259	1.1212	1.1165	1.1120
460	1.1652	1.1593	1.1537	1.1483	1.1430	1.1379	1.1330	1.1282
480	1.1843	1.1780	1.1720	1.1662	1.1606	1.1551	1.1499	1.1448
500	1.2040	1.1973	1.1909	1.1846	1.1786	1.1729	1.1672	1.1618
520	1.2243	1.2171	1.2103	1.2036	1.1972	1.1910	1.1851	1.1793
540	1.2452	1.2375	1.2302	1.2231	1.2163	1.2097	1.2034	1.1972
560	1.2666	1.2584	1.2506	1.2430	1.2358	1.2288	1.2221	1.2155
580	1.2886	1.2799	1.2715	1.2635	1.2557	1.2483	1.2412	1.2343
600	1.3111	1.3018	1.2929	1.2843	1.2761	1.2683	1.2607	1.2534
620	1.3341	1.3242	1.3147	1.3057	1.2970	1.2886	1.2806	1.2729
640	1.3576	1.3471	1.3371	1.3274	1.3182	1.3094	1.3009	1.2927
660	1.3817	1.3705	1.3598	1.3497	1.3399	1.3306	1.3216	1.3129
680	1.4062	1.3944	1.3831	1.3723	1.3620	1.3521	1.3427	1.3336
700	1.4313	1.4188	1.4068	1.3954	1.3846	1.3741	1.3641	1.3546
720	1.4569	1.4437	1.4311	1.4190	1.4075	1.3965	1.3860	1.3759
740	1.4830	1.4690	1.4557	1.4430	1.4309	1.4194	1.4083	1.3977
760	1.5097	1.4949	1.4809	1.4675	1.4548	1.4426	1.4310	1.4198
780	1.5367	1.5212	1.5064	1.4924	1.4790	1.4662	1.4540	1.4423
800	1.5642	1.5479	1.5324	1.5176	1.5035	1.4901	1.4773	1.4650
820	1.5921	1.5749	1.5586	1.5431	1.5284	1.5143	1.5009	1.4880
840	1.6202	1.6022	1.5851	1.5688	1.5533	1.5386	1.5245	1.5111
860	1.6483	1.6294	1.6115	1.5945	1.5783	1.5629	1.5482	1.5342
880	1.6763	1.6565	1.6378	1.6200	1.6031	1.5870	1.5716	1.5570
900	1.7039	1.6832	1.6636	1.6450	1.6273	1.6106	1.5946	1.5793
920	1.7307	1.7091	1.6886	1.6692	1.6508	1.6333	1.6167	1.6009
940	1.7563	1.7337	1.7124	1.6922	1.6731	1.6549	1.6377	1.6213
960	1.7802	1.7567	1.7345	1.7135	1.6936	1.6748	1.6570	1.6401
980	1.8018	1.7773	1.7542	1.7325	1.7119	1.6925	1.6741	1.6567
1000	1.8204	1.7950	1.7711	1.7486	1.7273	1.7074	1.6885	1.6707

TABLE 1
SPECIFIC VOLUME IN CUBIC CENTIMETERS PER GRAM

TEMP	PRESSURE IN BARS							
	4900	5000	5100	5200	5300	5400	5500	5600
20	0.8724	0.8707	0.8691	0.8675	0.8659	0.8643	0.8628	0.8613
40	0.8806	0.8789	0.8772	0.8756	0.8739	0.8723	0.8708	0.8692
60	0.8888	0.8870	0.8853	0.8836	0.8820	0.8803	0.8787	0.8772
80	0.8970	0.8953	0.8936	0.8919	0.8902	0.8885	0.8869	0.8852
100	0.9056	0.9038	0.9020	0.9003	0.8986	0.8969	0.8952	0.8936
120	0.9144	0.9126	0.9108	0.9090	0.9073	0.9055	0.9038	0.9021
140	0.9235	0.9217	0.9198	0.9180	0.9162	0.9144	0.9127	0.9110
160	0.9330	0.9311	0.9292	0.9274	0.9255	0.9237	0.9219	0.9201
180	0.9429	0.9409	0.9389	0.9370	0.9351	0.9332	0.9313	0.9294
200	0.9531	0.9511	0.9490	0.9470	0.9450	0.9430	0.9410	0.9391
220	0.9637	0.9615	0.9594	0.9573	0.9552	0.9531	0.9511	0.9490
240	0.9747	0.9724	0.9701	0.9679	0.9657	0.9635	0.9614	0.9593
260	0.9860	0.9836	0.9812	0.9788	0.9765	0.9742	0.9720	0.9698
280	0.9978	0.9952	0.9927	0.9902	0.9877	0.9853	0.9829	0.9806
300	1.0099	1.0072	1.0045	1.0019	0.9993	0.9967	0.9942	0.9918
320	1.0225	1.0196	1.0167	1.0139	1.0112	1.0085	1.0058	1.0032
340	1.0355	1.0324	1.0294	1.0264	1.0235	1.0207	1.0179	1.0151
360	1.0490	1.0457	1.0425	1.0393	1.0362	1.0332	1.0302	1.0273
380	1.0630	1.0595	1.0560	1.0527	1.0494	1.0462	1.0430	1.0399
400	1.0774	1.0737	1.0700	1.0664	1.0629	1.0595	1.0562	1.0529
420	1.0923	1.0883	1.0844	1.0806	1.0769	1.0733	1.0698	1.0663
440	1.1077	1.1034	1.0993	1.0953	1.0913	1.0875	1.0837	1.0801
460	1.1235	1.1190	1.1146	1.1103	1.1061	1.1021	1.0981	1.0942
480	1.1398	1.1350	1.1303	1.1258	1.1214	1.1171	1.1129	1.1087
500	1.1566	1.1515	1.1465	1.1417	1.1370	1.1324	1.1280	1.1236
520	1.1737	1.1683	1.1630	1.1579	1.1530	1.1481	1.1434	1.1388
540	1.1913	1.1855	1.1800	1.1746	1.1693	1.1642	1.1592	1.1544
560	1.2093	1.2032	1.1973	1.1915	1.1860	1.1806	1.1753	1.1702
580	1.2276	1.2212	1.2149	1.2089	1.2030	1.1973	1.1918	1.1864
600	1.2463	1.2395	1.2329	1.2265	1.2203	1.2143	1.2085	1.2028
620	1.2654	1.2582	1.2512	1.2445	1.2380	1.2316	1.2255	1.2195
640	1.2848	1.2772	1.2699	1.2628	1.2559	1.2493	1.2428	1.2365
660	1.3046	1.2966	1.2889	1.2814	1.2742	1.2672	1.2604	1.2538
680	1.3248	1.3164	1.3082	1.3004	1.2928	1.2854	1.2783	1.2713
700	1.3453	1.3365	1.3279	1.3196	1.3117	1.3039	1.2964	1.2892
720	1.3662	1.3569	1.3479	1.3392	1.3309	1.3228	1.3149	1.3073
740	1.3875	1.3777	1.3683	1.3592	1.3504	1.3419	1.3337	1.3257
760	1.4091	1.3988	1.3889	1.3794	1.3702	1.3613	1.3527	1.3444
780	1.4311	1.4203	1.4099	1.3999	1.3903	1.3810	1.3720	1.3633
800	1.4533	1.4420	1.4311	1.4207	1.4106	1.4009	1.3915	1.3824
820	1.4757	1.4639	1.4526	1.4417	1.4312	1.4210	1.4112	1.4018
840	1.4982	1.4859	1.4741	1.4627	1.4517	1.4412	1.4310	1.4211
860	1.5207	1.5079	1.4955	1.4837	1.4723	1.4613	1.4507	1.4405
880	1.5430	1.5296	1.5168	1.5045	1.4926	1.4813	1.4703	1.4597
900	1.5648	1.5509	1.5376	1.5248	1.5126	1.5008	1.4894	1.4785
920	1.5858	1.5714	1.5576	1.5445	1.5318	1.5197	1.5080	1.4968
940	1.6057	1.5908	1.5766	1.5631	1.5501	1.5377	1.5257	1.5142
960	1.6240	1.6087	1.5942	1.5803	1.5670	1.5544	1.5422	1.5306
980	1.6403	1.6246	1.6098	1.5957	1.5822	1.5694	1.5572	1.5455
1000	1.6539	1.6380	1.6230	1.6087	1.5952	1.5824	1.5702	1.5585

TABLE 1
SPECIFIC VOLUME IN CUBIC CENTIMETERS PER GRAM

TEMP	PRESSURE IN BARS							
	5700	5800	5900	6000	6100	6200	6300	6400
20	0.8598	0.8584	0.8570	0.8556	0.8543	0.8530	0.8517	0.8504
40	0.8677	0.8662	0.8648	0.8633	0.8619	0.8605	0.8592	0.8579
60	0.8756	0.8741	0.8726	0.8711	0.8696	0.8682	0.8668	0.8655
80	0.8837	0.8821	0.8805	0.8790	0.8775	0.8761	0.8747	0.8732
100	0.8919	0.8903	0.8887	0.8872	0.8857	0.8842	0.8827	0.8812
120	0.9005	0.8988	0.8972	0.8956	0.8940	0.8925	0.8910	0.8895
140	0.9092	0.9076	0.9059	0.9042	0.9026	0.9010	0.8995	0.8979
160	0.9183	0.9166	0.9148	0.9131	0.9115	0.9098	0.9082	0.9066
180	0.9276	0.9258	0.9240	0.9223	0.9205	0.9188	0.9171	0.9155
200	0.9372	0.9353	0.9335	0.9316	0.9298	0.9281	0.9263	0.9246
220	0.9471	0.9451	0.9432	0.9413	0.9394	0.9375	0.9357	0.9339
240	0.9572	0.9551	0.9531	0.9511	0.9492	0.9472	0.9453	0.9435
260	0.9676	0.9654	0.9633	0.9612	0.9592	0.9572	0.9552	0.9532
280	0.9783	0.9761	0.9738	0.9716	0.9695	0.9674	0.9653	0.9632
300	0.9893	0.9870	0.9846	0.9823	0.9801	0.9778	0.9757	0.9735
320	1.0007	0.9982	0.9957	0.9933	0.9909	0.9886	0.9863	0.9840
340	1.0124	1.0098	1.0072	1.0046	1.0021	0.9996	0.9972	0.9949
360	1.0245	1.0217	1.0189	1.0162	1.0136	1.0110	1.0085	1.0060
380	1.0369	1.0340	1.0311	1.0282	1.0254	1.0227	1.0200	1.0174
400	1.0497	1.0466	1.0435	1.0405	1.0376	1.0347	1.0319	1.0291
420	1.0629	1.0596	1.0564	1.0532	1.0501	1.0470	1.0440	1.0411
440	1.0765	1.0730	1.0696	1.0662	1.0629	1.0597	1.0565	1.0534
460	1.0904	1.0867	1.0831	1.0796	1.0761	1.0727	1.0693	1.0661
480	1.1047	1.1008	1.0970	1.0932	1.0896	1.0860	1.0824	1.0790
500	1.1194	1.1152	1.1112	1.1072	1.1033	1.0995	1.0958	1.0922
520	1.1343	1.1300	1.1257	1.1215	1.1174	1.1134	1.1095	1.1057
540	1.1496	1.1450	1.1405	1.1361	1.1318	1.1276	1.1234	1.1194
560	1.1652	1.1604	1.1556	1.1510	1.1464	1.1420	1.1376	1.1334
580	1.1811	1.1760	1.1710	1.1661	1.1613	1.1566	1.1521	1.1476
600	1.1973	1.1919	1.1866	1.1815	1.1764	1.1715	1.1667	1.1620
620	1.2137	1.2080	1.2025	1.1971	1.1918	1.1867	1.1816	1.1767
640	1.2304	1.2244	1.2186	1.2130	1.2074	1.2020	1.1967	1.1916
660	1.2474	1.2411	1.2350	1.2291	1.2233	1.2176	1.2121	1.2067
680	1.2646	1.2580	1.2517	1.2454	1.2394	1.2334	1.2277	1.2220
700	1.2821	1.2752	1.2686	1.2621	1.2557	1.2495	1.2434	1.2375
720	1.2999	1.2927	1.2857	1.2789	1.2723	1.2658	1.2595	1.2533
740	1.3180	1.3104	1.3031	1.2960	1.2891	1.2823	1.2757	1.2693
760	1.3363	1.3284	1.3208	1.3134	1.3061	1.2991	1.2922	1.2854
780	1.3549	1.3467	1.3387	1.3309	1.3234	1.3160	1.3088	1.3018
800	1.3736	1.3651	1.3568	1.3487	1.3408	1.3332	1.3257	1.3184
820	1.3926	1.3837	1.3750	1.3666	1.3584	1.3505	1.3427	1.3351
840	1.4116	1.4023	1.3934	1.3846	1.3761	1.3679	1.3598	1.3520
860	1.4306	1.4210	1.4117	1.4027	1.3939	1.3853	1.3770	1.3689
880	1.4494	1.4395	1.4299	1.4206	1.4115	1.4027	1.3941	1.3858
900	1.4680	1.4577	1.4479	1.4383	1.4290	1.4199	1.4111	1.4025
920	1.4860	1.4755	1.4654	1.4556	1.4461	1.4368	1.4279	1.4191
940	1.5032	1.4925	1.4822	1.4723	1.4627	1.4533	1.4442	1.4354
960	1.5194	1.5086	1.4983	1.4882	1.4786	1.4692	1.4601	1.4513
980	1.5342	1.5235	1.5131	1.5032	1.4936	1.4843	1.4753	1.4666
1000	1.5474	1.5368	1.5266	1.5169	1.5075	1.4985	1.4898	1.4814

TABLE 1
SPECIFIC VOLUME IN CUBIC CENTIMETERS PER GRAM

	PRESSURE IN BARS							
TEMP	6500	6600	6700	6800	6900	7000	7100	7200
20	0.8491	0.8479	0.8467	0.8454	0.8442	0.8431	0.8419	0.8407
40	0.8566	0.8553	0.8540	0.8528	0.8515	0.8503	0.8491	0.8479
60	0.8641	0.8628	0.8615	0.8602	0.8590	0.8577	0.8565	0.8553
80	0.8719	0.8705	0.8692	0.8678	0.8665	0.8652	0.8640	0.8627
100	0.8798	0.8784	0.8770	0.8757	0.8743	0.8730	0.8717	0.8704
120	0.8880	0.8865	0.8851	0.8837	0.8823	0.8809	0.8796	0.8783
140	0.8964	0.8949	0.8934	0.8919	0.8905	0.8891	0.8877	0.8863
160	0.9050	0.9034	0.9019	0.9004	0.8989	0.8975	0.8960	0.8946
180	0.9138	0.9122	0.9106	0.9091	0.9075	0.9060	0.9045	0.9030
200	0.9229	0.9212	0.9196	0.9179	0.9163	0.9147	0.9132	0.9116
220	0.9321	0.9304	0.9287	0.9270	0.9253	0.9237	0.9220	0.9204
240	0.9416	0.9398	0.9380	0.9362	0.9345	0.9328	0.9311	0.9294
260	0.9513	0.9494	0.9475	0.9457	0.9439	0.9421	0.9403	0.9386
280	0.9612	0.9592	0.9573	0.9554	0.9535	0.9516	0.9498	0.9479
300	0.9714	0.9693	0.9673	0.9652	0.9633	0.9613	0.9594	0.9575
320	0.9818	0.9796	0.9775	0.9754	0.9733	0.9713	0.9692	0.9673
340	0.9925	0.9902	0.9880	0.9857	0.9836	0.9814	0.9793	0.9772
360	1.0035	1.0011	0.9987	0.9964	0.9941	0.9918	0.9896	0.9874
380	1.0148	1.0122	1.0097	1.0073	1.0049	1.0025	1.0002	0.9979
400	1.0263	1.0237	1.0210	1.0184	1.0159	1.0134	1.0110	1.0085
420	1.0382	1.0354	1.0326	1.0299	1.0272	1.0246	1.0220	1.0194
440	1.0504	1.0474	1.0445	1.0416	1.0388	1.0360	1.0333	1.0306
460	1.0629	1.0597	1.0566	1.0536	1.0506	1.0477	1.0448	1.0420
480	1.0756	1.0723	1.0690	1.0658	1.0627	1.0596	1.0566	1.0536
500	1.0886	1.0851	1.0817	1.0783	1.0750	1.0718	1.0686	1.0654
520	1.1019	1.0982	1.0946	1.0911	1.0876	1.0842	1.0808	1.0775
540	1.1154	1.1116	1.1077	1.1040	1.1004	1.0968	1.0932	1.0898
560	1.1292	1.1251	1.1211	1.1172	1.1134	1.1096	1.1059	1.1022
580	1.1432	1.1389	1.1347	1.1306	1.1266	1.1226	1.1187	1.1149
600	1.1574	1.1529	1.1485	1.1442	1.1400	1.1358	1.1317	1.1277
620	1.1719	1.1672	1.1625	1.1580	1.1536	1.1492	1.1449	1.1407
640	1.1865	1.1816	1.1767	1.1720	1.1673	1.1628	1.1583	1.1539
660	1.2014	1.1962	1.1911	1.1862	1.1813	1.1765	1.1719	1.1673
680	1.2165	1.2110	1.2057	1.2006	1.1955	1.1905	1.1856	1.1808
700	1.2317	1.2261	1.2205	1.2151	1.2098	1.2046	1.1995	1.1945
720	1.2472	1.2413	1.2355	1.2299	1.2243	1.2189	1.2136	1.2084
740	1.2629	1.2568	1.2507	1.2448	1.2391	1.2334	1.2278	1.2224
760	1.2789	1.2724	1.2661	1.2600	1.2540	1.2481	1.2423	1.2366
780	1.2950	1.2883	1.2817	1.2753	1.2690	1.2629	1.2569	1.2510
800	1.3113	1.3043	1.2975	1.2908	1.2843	1.2779	1.2717	1.2655
820	1.3277	1.3205	1.3134	1.3065	1.2997	1.2931	1.2866	1.2802
840	1.3443	1.3368	1.3295	1.3223	1.3153	1.3084	1.3017	1.2951
860	1.3609	1.3532	1.3456	1.3382	1.3310	1.3239	1.3170	1.3102
880	1.3776	1.3696	1.3618	1.3542	1.3468	1.3395	1.3323	1.3253
900	1.3942	1.3860	1.3780	1.3703	1.3626	1.3552	1.3479	1.3407
920	1.4106	1.4023	1.3942	1.3863	1.3785	1.3710	1.3635	1.3562
940	1.4268	1.4184	1.4103	1.4023	1.3945	1.3868	1.3794	1.3720
960	1.4427	1.4343	1.4262	1.4183	1.4105	1.4029	1.3954	1.3881
980	1.4582	1.4500	1.4420	1.4342	1.4266	1.4191	1.4118	1.4047
1000	1.4732	1.4653	1.4576	1.4501	1.4428	1.4357	1.4287	1.4219

TABLE 1
SPECIFIC VOLUME IN CUBIC CENTIMETERS PER GRAM

TEMP	PRESSURE IN BARS							
	7300	7400	7500	7600	7700	7800	7900	8000
20	0.8395	0.8383	0.8371	0.8359	0.8347	0.8335	0.8323	0.8311
40	0.8467	0.8455	0.8444	0.8432	0.8420	0.8408	0.8397	0.8385
60	0.8540	0.8528	0.8516	0.8505	0.8493	0.8481	0.8469	0.8458
80	0.8615	0.8603	0.8591	0.8578	0.8566	0.8555	0.8543	0.8531
100	0.8691	0.8679	0.8666	0.8654	0.8642	0.8629	0.8617	0.8605
120	0.8769	0.8756	0.8744	0.8731	0.8718	0.8706	0.8693	0.8681
140	0.8850	0.8836	0.8823	0.8810	0.8797	0.8784	0.8771	0.8758
160	0.8932	0.8918	0.8904	0.8890	0.8877	0.8863	0.8850	0.8837
180	0.9016	0.9001	0.8987	0.8973	0.8959	0.8945	0.8931	0.8917
200	0.9101	0.9086	0.9071	0.9057	0.9042	0.9028	0.9013	0.8999
220	0.9189	0.9173	0.9158	0.9142	0.9127	0.9112	0.9097	0.9083
240	0.9278	0.9262	0.9245	0.9230	0.9214	0.9198	0.9183	0.9168
260	0.9369	0.9352	0.9335	0.9319	0.9302	0.9286	0.9270	0.9254
280	0.9462	0.9444	0.9426	0.9409	0.9392	0.9375	0.9359	0.9342
300	0.9556	0.9538	0.9520	0.9502	0.9484	0.9466	0.9449	0.9432
320	0.9653	0.9634	0.9615	0.9596	0.9577	0.9559	0.9541	0.9523
340	0.9752	0.9732	0.9712	0.9692	0.9673	0.9653	0.9635	0.9616
360	0.9853	0.9832	0.9811	0.9790	0.9770	0.9750	0.9730	0.9710
380	0.9956	0.9934	0.9912	0.9890	0.9869	0.9848	0.9827	0.9806
400	1.0062	1.0038	1.0015	0.9992	0.9970	0.9948	0.9926	0.9904
420	1.0169	1.0145	1.0121	1.0097	1.0073	1.0050	1.0027	1.0004
440	1.0280	1.0254	1.0228	1.0203	1.0178	1.0154	1.0130	1.0106
460	1.0392	1.0365	1.0338	1.0311	1.0285	1.0260	1.0234	1.0209
480	1.0507	1.0478	1.0450	1.0422	1.0394	1.0367	1.0341	1.0314
500	1.0624	1.0593	1.0564	1.0534	1.0505	1.0477	1.0449	1.0421
520	1.0743	1.0711	1.0679	1.0649	1.0618	1.0588	1.0559	1.0530
540	1.0864	1.0830	1.0797	1.0765	1.0733	1.0701	1.0670	1.0640
560	1.0987	1.0951	1.0917	1.0883	1.0849	1.0816	1.0784	1.0752
580	1.1111	1.1074	1.1038	1.1002	1.0967	1.0933	1.0899	1.0865
600	1.1238	1.1199	1.1161	1.1124	1.1087	1.1051	1.1015	1.0980
620	1.1366	1.1326	1.1286	1.1247	1.1208	1.1170	1.1133	1.1096
640	1.1496	1.1454	1.1412	1.1371	1.1331	1.1292	1.1253	1.1214
660	1.1628	1.1583	1.1540	1.1497	1.1455	1.1414	1.1374	1.1334
680	1.1761	1.1715	1.1669	1.1625	1.1581	1.1538	1.1496	1.1454
700	1.1896	1.1848	1.1800	1.1754	1.1708	1.1664	1.1619	1.1576
720	1.2032	1.1982	1.1933	1.1885	1.1837	1.1790	1.1744	1.1699
740	1.2171	1.2118	1.2067	1.2017	1.1967	1.1919	1.1871	1.1824
760	1.2311	1.2256	1.2203	1.2150	1.2099	1.2048	1.1998	1.1950
780	1.2452	1.2395	1.2340	1.2285	1.2232	1.2179	1.2128	1.2077
800	1.2595	1.2536	1.2479	1.2422	1.2366	1.2312	1.2258	1.2205
820	1.2740	1.2679	1.2619	1.2560	1.2503	1.2446	1.2390	1.2336
840	1.2887	1.2823	1.2761	1.2700	1.2641	1.2582	1.2524	1.2468
860	1.3035	1.2969	1.2905	1.2842	1.2781	1.2720	1.2661	1.2602
880	1.3185	1.3118	1.3052	1.2987	1.2923	1.2861	1.2799	1.2739
900	1.3337	1.3268	1.3200	1.3134	1.3069	1.3005	1.2942	1.2880
920	1.3491	1.3421	1.3352	1.3285	1.3219	1.3153	1.3089	1.3026
940	1.3648	1.3578	1.3509	1.3441	1.3374	1.3308	1.3243	1.3179
960	1.3810	1.3740	1.3671	1.3603	1.3536	1.3470	1.3405	1.3341
980	1.3977	1.3908	1.3840	1.3774	1.3708	1.3644	1.3580	1.3517
1000	1.4152	1.4086	1.4021	1.3957	1.3894	1.3832	1.3770	1.3709

TABLE 1
SPECIFIC VOLUME IN CUBIC CENTIMETERS PER GRAM

TEMP	PRESSURE IN BARS 8100	8200	8300	8400	8500	8600	8700	8800
20	0.8299	0.8287	0.8275	0.8262	0.8250	0.8238	0.8226	0.8214
40	0.8373	0.8361	0.8350	0.8338	0.8326	0.8315	0.8303	0.8292
60	0.8446	0.8435	0.8423	0.8412	0.8400	0.8389	0.8378	0.8367
80	0.8519	0.8508	0.8496	0.8485	0.8473	0.8462	0.8451	0.8440
100	0.8593	0.8582	0.8570	0.8558	0.8547	0.8535	0.8524	0.8513
120	0.8669	0.8657	0.8645	0.8633	0.8621	0.8609	0.8598	0.8586
140	0.8746	0.8733	0.8721	0.8708	0.8696	0.8684	0.8672	0.8660
160	0.8824	0.8811	0.8798	0.8785	0.8773	0.8760	0.8748	0.8735
180	0.8904	0.8890	0.8877	0.8864	0.8851	0.8838	0.8825	0.8812
200	0.8985	0.8971	0.8958	0.8944	0.8930	0.8917	0.8903	0.8890
220	0.9068	0.9054	0.9039	0.9025	0.9011	0.8997	0.8983	0.8969
240	0.9153	0.9138	0.9123	0.9108	0.9093	0.9079	0.9064	0.9050
260	0.9239	0.9223	0.9207	0.9192	0.9177	0.9162	0.9147	0.9132
280	0.9326	0.9310	0.9294	0.9278	0.9262	0.9246	0.9230	0.9215
300	0.9415	0.9398	0.9381	0.9365	0.9348	0.9332	0.9315	0.9299
320	0.9505	0.9488	0.9470	0.9453	0.9436	0.9419	0.9402	0.9385
340	0.9597	0.9579	0.9561	0.9543	0.9525	0.9507	0.9489	0.9472
360	0.9691	0.9672	0.9653	0.9634	0.9615	0.9596	0.9578	0.9559
380	0.9786	0.9766	0.9746	0.9726	0.9707	0.9687	0.9668	0.9649
400	0.9883	0.9862	0.9841	0.9820	0.9800	0.9779	0.9759	0.9739
420	0.9982	0.9960	0.9938	0.9916	0.9894	0.9873	0.9851	0.9830
440	1.0082	1.0059	1.0036	1.0013	0.9990	0.9968	0.9945	0.9923
460	1.0184	1.0160	1.0135	1.0111	1.0088	1.0064	1.0040	1.0017
480	1.0288	1.0262	1.0237	1.0211	1.0186	1.0162	1.0137	1.0112
500	1.0394	1.0367	1.0340	1.0313	1.0287	1.0261	1.0235	1.0209
520	1.0501	1.0472	1.0444	1.0416	1.0389	1.0361	1.0334	1.0307
540	1.0610	1.0580	1.0550	1.0521	1.0492	1.0463	1.0435	1.0406
560	1.0720	1.0689	1.0658	1.0627	1.0597	1.0567	1.0537	1.0507
580	1.0832	1.0799	1.0767	1.0735	1.0703	1.0671	1.0640	1.0609
600	1.0945	1.0911	1.0877	1.0844	1.0811	1.0778	1.0745	1.0713
620	1.1060	1.1024	1.0989	1.0954	1.0920	1.0885	1.0851	1.0818
640	1.1176	1.1139	1.1102	1.1066	1.1030	1.0994	1.0959	1.0924
660	1.1294	1.1255	1.1217	1.1179	1.1141	1.1104	1.1068	1.1031
680	1.1413	1.1373	1.1333	1.1293	1.1254	1.1216	1.1178	1.1140
700	1.1533	1.1491	1.1450	1.1409	1.1368	1.1328	1.1289	1.1250
720	1.1655	1.1611	1.1568	1.1525	1.1483	1.1442	1.1401	1.1361
740	1.1778	1.1732	1.1687	1.1643	1.1600	1.1557	1.1514	1.1472
760	1.1902	1.1854	1.1808	1.1762	1.1717	1.1673	1.1629	1.1585
780	1.2027	1.1978	1.1930	1.1882	1.1835	1.1789	1.1744	1.1699
800	1.2154	1.2103	1.2053	1.2004	1.1955	1.1907	1.1860	1.1814
820	1.2282	1.2229	1.2177	1.2126	1.2076	1.2027	1.1978	1.1930
840	1.2412	1.2358	1.2304	1.2251	1.2199	1.2148	1.2097	1.2047
860	1.2545	1.2488	1.2433	1.2378	1.2324	1.2271	1.2219	1.2167
880	1.2680	1.2622	1.2564	1.2508	1.2452	1.2398	1.2344	1.2290
900	1.2819	1.2760	1.2701	1.2643	1.2585	1.2529	1.2473	1.2418
920	1.2964	1.2903	1.2843	1.2783	1.2725	1.2667	1.2610	1.2553
940	1.3116	1.3054	1.2993	1.2933	1.2873	1.2814	1.2756	1.2698
960	1.3278	1.3216	1.3155	1.3094	1.3034	1.2974	1.2915	1.2856
980	1.3454	1.3393	1.3332	1.3271	1.3211	1.3152	1.3093	1.3034
1000	1.3649	1.3589	1.3530	1.3471	1.3412	1.3353	1.3295	1.3237

TABLE 1
SPECIFIC VOLUME IN CUBIC CENTIMETERS PER GRAM

TEMP	PRESSURE IN BARS 8900	9000	9100	9200	9300	9400	9500	9600
20	(0.8202)	(0.8190)	(0.8179)	(0.8168)	(0.8157)	(0.8146)	(0.8135)	(0.8125)
40	0.8281	0.8270	0.8259	0.8249	0.8238	0.8228	0.8218	0.8209
60	0.8356	0.8345	0.8335	0.8325	0.8314	0.8305	0.8295	0.8285
80	0.8429	0.8419	0.8408	0.8398	0.8387	0.8377	0.8368	0.8358
100	0.8502	0.8491	0.8480	0.8470	0.8459	0.8449	0.8439	0.8429
120	0.8575	0.8563	0.8552	0.8541	0.8530	0.8520	0.8509	0.8499
140	0.8648	0.8637	0.8625	0.8614	0.8602	0.8591	0.8580	0.8569
160	0.8723	0.8711	0.8699	0.8687	0.8675	0.8663	0.8652	0.8640
180	0.8799	0.8786	0.8774	0.8761	0.8749	0.8737	0.8725	0.8713
200	0.8877	0.8863	0.8850	0.8837	0.8824	0.8812	0.8799	0.8786
220	0.8955	0.8942	0.8928	0.8914	0.8901	0.8888	0.8874	0.8861
240	0.9035	0.9021	0.9007	0.8993	0.8979	0.8965	0.8951	0.8938
260	0.9117	0.9102	0.9087	0.9073	0.9058	0.9044	0.9030	0.9015
280	0.9199	0.9184	0.9169	0.9154	0.9139	0.9124	0.9109	0.9094
300	0.9283	0.9267	0.9251	0.9236	0.9220	0.9204	0.9189	0.9174
320	0.9368	0.9351	0.9335	0.9319	0.9302	0.9286	0.9270	0.9254
340	0.9454	0.9437	0.9419	0.9402	0.9385	0.9369	0.9352	0.9335
360	0.9541	0.9523	0.9505	0.9487	0.9469	0.9452	0.9434	0.9417
380	0.9629	0.9610	0.9592	0.9573	0.9554	0.9536	0.9518	0.9500
400	0.9719	0.9699	0.9679	0.9660	0.9640	0.9621	0.9602	0.9583
420	0.9809	0.9788	0.9768	0.9747	0.9727	0.9706	0.9686	0.9666
440	0.9901	0.9879	0.9857	0.9836	0.9814	0.9793	0.9772	0.9751
460	0.9994	0.9971	0.9948	0.9925	0.9903	0.9880	0.9858	0.9836
480	1.0088	1.0064	1.0040	1.0016	0.9992	0.9968	0.9945	0.9922
500	1.0183	1.0158	1.0133	1.0108	1.0083	1.0058	1.0033	1.0009
520	1.0280	1.0253	1.0227	1.0201	1.0174	1.0148	1.0123	1.0097
540	1.0378	1.0350	1.0322	1.0295	1.0267	1.0240	1.0213	1.0186
560	1.0478	1.0448	1.0419	1.0390	1.0362	1.0333	1.0305	1.0277
580	1.0578	1.0548	1.0518	1.0487	1.0457	1.0428	1.0398	1.0369
600	1.0681	1.0649	1.0617	1.0586	1.0554	1.0523	1.0492	1.0462
620	1.0784	1.0751	1.0718	1.0685	1.0653	1.0620	1.0588	1.0557
640	1.0889	1.0855	1.0820	1.0786	1.0753	1.0719	1.0686	1.0653
660	1.0995	1.0959	1.0924	1.0889	1.0854	1.0819	1.0784	1.0750
680	1.1103	1.1065	1.1029	1.0992	1.0956	1.0920	1.0884	1.0849
700	1.1211	1.1173	1.1135	1.1097	1.1059	1.1022	1.0985	1.0949
720	1.1320	1.1281	1.1241	1.1202	1.1164	1.1125	1.1087	1.1050
740	1.1431	1.1390	1.1349	1.1309	1.1269	1.1229	1.1190	1.1151
760	1.1542	1.1500	1.1458	1.1416	1.1375	1.1334	1.1294	1.1254
780	1.1654	1.1611	1.1567	1.1524	1.1482	1.1440	1.1398	1.1357
800	1.1768	1.1722	1.1677	1.1633	1.1589	1.1546	1.1503	1.1460
820	1.1882	1.1835	1.1789	1.1743	1.1697	1.1653	1.1608	1.1564
840	1.1998	1.1949	1.1901	1.1854	1.1807	1.1761	1.1715	1.1669
860	1.2116	1.2066	1.2016	1.1967	1.1919	1.1871	1.1823	1.1776
880	1.2238	1.2186	1.2134	1.2084	1.2033	1.1984	1.1934	1.1886
900	1.2364	1.2310	1.2257	1.2205	1.2153	1.2101	1.2050	1.2000
920	1.2497	1.2442	1.2388	1.2333	1.2280	1.2226	1.2174	1.2121
940	1.2641	1.2584	1.2528	1.2472	1.2417	1.2362	1.2307	1.2253
960	1.2798	1.2741	1.2683	1.2626	1.2569	1.2513	1.2457	1.2401
980	1.2975	1.2917	1.2859	1.2801	1.2743	1.2685	1.2628	1.2570
1000	1.3179	1.3121	1.3062	1.3004	1.2946	1.2887	1.2828	1.2769

TABLE 1
SPECIFIC VOLUME IN CUBIC CENTIMETERS PER GRAM

TEMP	PRESSURE IN BARS 9700	9800	9900	10000
20	(0.8115)	(0.8105)	(0.8096)	(0.8086)
40	0.8199	0.8190	0.8181	0.8172
60	0.8276	0.8267	0.8258	0.8248
80	0.8349	0.8339	0.8330	0.8320
100	0.8419	0.8409	0.8399	0.8389
120	0.8488	0.8478	0.8468	0.8457
140	0.8558	0.8547	0.8536	0.8525
160	0.8629	0.8617	0.8606	0.8594
180	0.8700	0.8688	0.8676	0.8664
200	0.8774	0.8761	0.8749	0.8736
220	0.8848	0.8835	0.8822	0.8809
240	0.8924	0.8911	0.8897	0.8884
260	0.9001	0.8987	0.8974	0.8960
280	0.9080	0.9065	0.9051	0.9037
300	0.9159	0.9144	0.9129	0.9114
320	0.9239	0.9223	0.9208	0.9193
340	0.9319	0.9303	0.9287	0.9272
360	0.9400	0.9383	0.9367	0.9351
380	0.9482	0.9464	0.9447	0.9430
400	0.9564	0.9546	0.9528	0.9510
420	0.9647	0.9628	0.9609	0.9590
440	0.9730	0.9710	0.9690	0.9671
460	0.9814	0.9793	0.9772	0.9752
480	0.9899	0.9877	0.9855	0.9833
500	0.9985	0.9961	0.9938	0.9915
520	1.0072	1.0047	1.0022	0.9998
540	1.0160	1.0134	1.0108	1.0083
560	1.0249	1.0222	1.0194	1.0168
580	1.0340	1.0311	1.0283	1.0255
600	1.0432	1.0402	1.0372	1.0343
620	1.0525	1.0494	1.0463	1.0433
640	1.0620	1.0588	1.0556	1.0524
660	1.0716	1.0683	1.0650	1.0617
680	1.0814	1.0779	1.0745	1.0711
700	1.0913	1.0877	1.0841	1.0807
720	1.1012	1.0976	1.0939	1.0903
740	1.1113	1.1075	1.1037	1.1001
760	1.1214	1.1175	1.1136	1.1098
780	1.1316	1.1276	1.1236	1.1197
800	1.1418	1.1377	1.1336	1.1295
820	1.1521	1.1478	1.1436	1.1394
840	1.1624	1.1580	1.1536	1.1493
860	1.1730	1.1684	1.1638	1.1594
880	1.1838	1.1790	1.1743	1.1696
900	1.1950	1.1900	1.1851	1.1803
920	1.2069	1.2018	1.1967	1.1916
940	1.2199	1.2146	1.2093	1.2040
960	1.2345	1.2289	1.2234	1.2179
980	1.2512	1.2455	1.2397	1.2340
1000	1.2710	1.2651	1.2592	1.2532

TABLE 2
GIBBS FREE ENERGY IN CALORIES PER MOLE

TEMP	PRESSURE IN BARS							
	100	200	300	400	500	600	700	800
20	30	73	115	158	201	242	284	326
40	-6	36	79	122	165	207	249	291
60	-66	-24	20	63	106	149	191	234
80	-148	-103	-60	-16	26	70	113	155
100	-249	-205	-162	-116	-72	-30	13	57
120	-372	-326	-280	-236	-192	-146	-103	-58
140	-511	-465	-419	-374	-328	-283	-237	-193
160	-668	-622	-574	-529	-483	-437	-391	-344
180	-843	-794	-747	-701	-652	-605	-559	-513
200	-1035	-985	-936	-888	-839	-792	-744	-697
220	-1242	-1191	-1141	-1091	-1043	-993	-944	-897
240	-1466	-1414	-1363	-1310	-1259	-1210	-1161	-1111
260	-1706	-1651	-1597	-1544	-1493	-1441	-1390	-1338
280	-1959	-1903	-1848	-1794	-1739	-1688	-1634	-1582
300	-2231	-2169	-2113	-2054	-1999	-1946	-1888	-1837
320	-2601	-2455	-2391	-2337	-2276	-2221	-2163	-2104
340	-3100	-2755	-2691	-2629	-2564	-2505	-2447	-2394
360	-3616	-3077	-3002	-2938	-2872	-2809	-2745	-2684
380	-4140	-3492	-3337	-3260	-3188	-3123	-3062	-2997
400	-4669	-3962	-3700	-3605	-3526	-3457	-3385	-3318
420	-5205	-4449	-4112	-3972	-3882	-3800	-3725	-3657
440	-5755	-4945	-4561	-4369	-4252	-4161	-4081	-4009
460	-6303	-5452	-5031	-4792	-4651	-4546	-4453	-4373
480	-6861	-5966	-5513	-5242	-5065	-4944	-4843	-4753
500	-7429	-6495	-6007	-5706	-5506	-5360	-5245	-5145
520	-7998	-7024	-6507	-6182	-5959	-5792	-5661	-5553
540	-8577	-7563	-7021	-6669	-6424	-6238	-6095	-5973
560	-9158	-8109	-7544	-7169	-6902	-6700	-6538	-6405
580	-9744	-8662	-8069	-7673	-7386	-7171	-6997	-6851
600	-10337	-9219	-8598	-8188	-7889	-7656	-7465	-7305
620	-10931	-9781	-9140	-8712	-8388	-8144	-7940	-7775
640	-11532	-10349	-9687	-9234	-8899	-8639	-8428	-8251
660	-12137	-10920	-10238	-9771	-9423	-9147	-8923	-8729
680	-12751	-11497	-10793	-10307	-9948	-9658	-9425	-9225
700	-13362	-12079	-11354	-10855	-10477	-10179	-9935	-9723
720	-13982	-12667	-11921	-11405	-11018	-10708	-10450	-10228
740	-14606	-13259	-12493	-11963	-11559	-11241	-10968	-10747
760	-15229	-13856	-13071	-12525	-12109	-11777	-11499	-11261
780	-15860	-14453	-13653	-13089	-12660	-12324	-12032	-11786
800	-16499	-15058	-14235	-13662	-13223	-12867	-12571	-12318

TABLE 2
GIBBS FREE ENERGY IN CALORIES PER MOLE

TEMP	PRESSURE IN BARS 900	1000	1100	1200	1300	1400	1500	1600
20	367	409	450	491	532	573	614	655
40	333	375	416	458	500	541	582	623
60	275	318	360	402	443	485	527	568
80	198	241	284	326	368	410	452	494
100	100	144	187	230	273	315	358	400
120	-15	27	71	114	157	200	243	286
140	-149	-104	-60	-16	27	70	114	157
160	-299	-255	-210	-165	-121	-77	-32	11
180	-467	-422	-376	-331	-286	-241	-196	-151
200	-650	-604	-558	-511	-465	-419	-374	-328
220	-849	-800	-753	-706	-659	-612	-565	-519
240	-1062	-1013	-964	-916	-868	-821	-773	-726
260	-1289	-1239	-1190	-1141	-1092	-1043	-995	-947
280	-1531	-1479	-1428	-1378	-1328	-1278	-1229	-1180
300	-1786	-1733	-1681	-1629	-1578	-1527	-1477	-1427
320	-2048	-1997	-1943	-1890	-1838	-1786	-1734	-1683
340	-2335	-2278	-2223	-2168	-2115	-2061	-2008	-1956
360	-2625	-2570	-2513	-2456	-2401	-2346	-2292	-2238
380	-2935	-2875	-2816	-2758	-2700	-2644	-2588	-2533
400	-3253	-3191	-3129	-3069	-3010	-2951	-2894	-2837
420	-3591	-3520	-3456	-3394	-3332	-3272	-3212	-3153
440	-3937	-3866	-3799	-3733	-3669	-3607	-3545	-3485
460	-4296	-4221	-4150	-4082	-4015	-3950	-3886	-3823
480	-4670	-4589	-4514	-4442	-4372	-4305	-4238	-4173
500	-5054	-4968	-4888	-4813	-4739	-4668	-4599	-4532
520	-5455	-5361	-5276	-5196	-5119	-5044	-4972	-4902
540	-5864	-5769	-5679	-5594	-5513	-5435	-5359	-5286
560	-6291	-6183	-6087	-5997	-5911	-5829	-5750	-5674
580	-6725	-6612	-6509	-6413	-6323	-6237	-6154	-6075
600	-7169	-7053	-6943	-6841	-6746	-6656	-6569	-6486
620	-7625	-7500	-7383	-7276	-7175	-7080	-6990	-6903
640	-8094	-7957	-7833	-7719	-7613	-7514	-7419	-7328
660	-8570	-8421	-8290	-8170	-8058	-7954	-7855	-7760
680	-9052	-8898	-8760	-8634	-8517	-8407	-8304	-8205
700	-9545	-9380	-9235	-9102	-8980	-8865	-8757	-8655
720	-10038	-9873	-9721	-9582	-9454	-9335	-9222	-9116
740	-10544	-10371	-10213	-10068	-9934	-9810	-9693	-9582
760	-11059	-10870	-10705	-10554	-10415	-10286	-10165	-10050
780	-11572	-11386	-11215	-11058	-10914	-10780	-10654	-10534
800	-12098	-11901	-11723	-11560	-11411	-11272	-11141	-11018
820		-12424	-12240	-12072	-11917	-11773	-11638	-11511
840		-12952	-12762	-12588	-12428	-12280	-12140	-12009
860		-13487	-13291	-13112	-12947	-12793	-12650	-12514
880		-14026	-13824	-13639	-13469	-13311	-13163	-13023
900		-14570	-14361	-14172	-13997	-13834	-13682	-13539
920		-15120	-14906	-14711	-14532	-14365	-14209	-14062
940		-15673	-15454	-15255	-15071	-14900	-14740	-14590
960		-16232	-16008	-15804	-15616	-15441	-15278	-15123
980		-16794	-16566	-16357	-16165	-15986	-15818	-15660
1000		-17361	-17128	-16915	-16718	-16535	-16363	-16202

TABLE 2
GIBBS FREE ENERGY IN CALORIES PER MOLE

	PRESSURE IN BARS							
TEMP	1700	1800	1900	2000	2100	2200	2300	2400
20	695	736	776	816	856	896	936	976
40	664	705	745	786	826	867	907	947
60	609	650	691	732	773	814	854	895
80	536	577	619	660	701	742	783	824
100	442	484	526	568	609	651	692	733
120	329	371	413	456	498	540	582	623
140	201	244	286	329	372	414	457	499
160	55	98	142	185	228	271	314	357
180	-106	-62	-18	25	69	113	156	200
200	-283	-238	-193	-148	-103	-59	-15	28
220	-473	-427	-381	-336	-291	-245	-201	-156
240	-679	-632	-586	-540	-493	-448	-402	-357
260	-899	-851	-804	-757	-710	-663	-617	-570
280	-1131	-1083	-1034	-986	-939	-891	-844	-797
300	-1377	-1327	-1278	-1229	-1180	-1132	-1084	-1036
320	-1632	-1582	-1531	-1481	-1432	-1382	-1333	-1285
340	-1904	-1852	-1801	-1750	-1699	-1649	-1599	-1549
360	-2185	-2132	-2079	-2027	-1975	-1924	-1873	-1822
380	-2478	-2424	-2370	-2316	-2264	-2211	-2159	-2107
400	-2781	-2725	-2670	-2615	-2561	-2507	-2454	-2401
420	-3095	-3038	-2981	-2925	-2870	-2815	-2760	-2706
440	-3425	-3366	-3308	-3250	-3192	-3136	-3080	-3025
460	-3762	-3701	-3641	-3582	-3522	-3464	-3407	-3350
480	-4109	-4046	-3985	-3924	-3863	-3803	-3745	-3686
500	-4466	-4401	-4337	-4274	-4212	-4151	-4090	-4031
520	-4833	-4766	-4700	-4635	-4572	-4509	-4447	-4386
540	-5215	-5145	-5077	-5010	-4946	-4881	-4817	-4754
560	-5600	-5527	-5457	-5388	-5323	-5256	-5191	-5126
580	-5998	-5922	-5849	-5778	-5712	-5644	-5576	-5509
600	-6406	-6328	-6252	-6178	-6112	-6041	-5971	-5903
620	-6819	-6738	-6660	-6583	-6515	-6442	-6370	-6300
640	-7241	-7157	-7075	-6996	-6927	-6851	-6777	-6705
660	-7669	-7582	-7497	-7415	-7344	-7266	-7189	-7115
680	-8111	-8020	-7933	-7848	-7773	-7693	-7614	-7537
700	-8557	-8463	-8372	-8284	-8206	-8123	-8042	-7963
720	-9014	-8916	-8822	-8731	-8649	-8564	-8480	-8399
740	-9477	-9375	-9278	-9184	-9098	-9010	-8924	-8841
760	-9940	-9836	-9735	-9638	-9547	-9456	-9368	-9282
780	-10421	-10313	-10209	-10109	-10014	-9920	-9829	-9741
800	-10901	-10789	-10682	-10579	-10479	-10383	-10289	-10199
820	-11390	-11275	-11164	-11058	-10954	-10855	-10759	-10666
840	-11884	-11765	-11652	-11542	-11435	-11333	-11235	-11139
860	-12386	-12263	-12146	-12034	-11924	-11820	-11719	-11621
880	-12891	-12765	-12645	-12530	-12418	-12311	-12207	-12107
900	-13403	-13274	-13151	-13033	-12919	-12809	-12703	-12600
920	-13923	-13791	-13664	-13543	-13426	-13314	-13205	-13099
940	-14447	-14311	-14182	-14057	-13938	-13822	-13711	-13603
960	-14977	-14838	-14705	-14578	-14456	-14338	-14224	-14113
980	-15511	-15368	-15232	-15102	-14977	-14856	-14739	-14627
1000	-16048	-15903	-15763	-15630	-15502	-15379	-15259	-15144

TABLE 2
GIBBS FREE ENERGY IN CALORIES PER MOLE

TEMP	PRESSURE IN BARS 2500	2600	2700	2800	2900	3000	3100	3200
20	1016	1055	1094	1134	1173	1212	1251	1290
40	987	1027	1067	1106	1146	1185	1225	1264
60	935	975	1015	1055	1095	1135	1175	1215
80	865	905	946	986	1026	1067	1107	1147
100	774	815	856	897	938	978	1019	1059
120	665	706	748	789	830	871	912	953
140	541	583	625	666	708	750	791	832
160	400	442	485	527	569	611	653	695
180	243	286	329	372	414	457	499	542
200	72	116	159	203	246	289	332	375
220	-111	-67	-23	20	64	108	152	195
240	-311	-266	-221	-177	-132	-88	-43	0
260	-524	-479	-433	-388	-342	-297	-252	-208
280	-750	-703	-657	-611	-565	-519	-474	-428
300	-988	-941	-894	-847	-800	-754	-707	-661
320	-1236	-1188	-1140	-1092	-1045	-997	-950	-903
340	-1500	-1450	-1402	-1353	-1305	-1257	-1209	-1161
360	-1772	-1722	-1672	-1622	-1573	-1524	-1475	-1427
380	-2056	-2005	-1954	-1903	-1853	-1803	-1754	-1705
400	-2348	-2296	-2244	-2193	-2142	-2091	-2041	-1991
420	-2652	-2599	-2546	-2494	-2442	-2390	-2338	-2287
440	-2970	-2915	-2862	-2808	-2755	-2702	-2650	-2598
460	-3294	-3239	-3183	-3129	-3074	-3021	-2967	-2914
480	-3629	-3572	-3516	-3460	-3404	-3349	-3295	-3240
500	-3972	-3913	-3856	-3799	-3742	-3686	-3630	-3575
520	-4326	-4266	-4207	-4148	-4090	-4033	-3976	-3920
540	-4692	-4631	-4570	-4511	-4451	-4393	-4335	-4277
560	-5062	-4999	-4937	-4876	-4815	-4755	-4696	-4637
580	-5444	-5379	-5316	-5253	-5191	-5129	-5069	-5009
600	-5835	-5769	-5704	-5639	-5576	-5513	-5451	-5389
620	-6231	-6163	-6096	-6030	-5965	-5900	-5837	-5774
640	-6634	-6564	-6495	-6427	-6361	-6295	-6230	-6166
660	-7042	-6970	-6899	-6830	-6762	-6694	-6628	-6563
680	-7462	-7389	-7316	-7245	-7175	-7106	-7038	-6971
700	-7886	-7810	-7736	-7663	-7591	-7521	-7451	-7383
720	-8320	-8242	-8166	-8091	-8018	-7945	-7874	-7804
740	-8759	-8679	-8601	-8524	-8449	-8375	-8303	-8231
760	-9198	-9116	-9036	-8958	-8881	-8805	-8731	-8658
780	-9655	-9571	-9489	-9408	-9329	-9252	-9176	-9101
800	-10110	-10024	-9940	-9857	-9777	-9697	-9620	-9543
820	-10576	-10487	-10401	-10316	-10234	-10153	-10073	-9995
840	-11046	-10956	-10867	-10781	-10696	-10613	-10532	-10452
860	-11525	-11432	-11342	-11253	-11167	-11082	-10999	-10917
880	-12009	-11914	-11821	-11731	-11642	-11556	-11471	-11388
900	-12500	-12402	-12307	-12215	-12124	-12036	-11949	-11864
920	-12997	-12897	-12800	-12706	-12613	-12523	-12435	-12348
940	-13498	-13397	-13297	-13201	-13106	-13014	-12924	-12836
960	-14006	-13902	-13801	-13702	-13606	-13512	-13420	-13330
980	-14517	-14411	-14308	-14207	-14109	-14014	-13920	-13828
1000	-15033	-14924	-14819	-14717	-14617	-14519	-14424	-14331

TABLE 2
GIBBS FREE ENERGY IN CALORIES PER MOLE

TEMP	PRESSURE IN BARS 3300	3400	3500	3600	3700	3800	3900	4000
20	1329	1368	1406	1445	1484	1522	1560	1599
40	1303	1342	1381	1420	1459	1498	1537	1575
60	1254	1294	1333	1372	1412	1451	1490	1529
80	1187	1226	1266	1306	1346	1385	1424	1464
100	1100	1140	1180	1220	1260	1300	1340	1379
120	994	1034	1075	1115	1156	1196	1236	1276
140	873	915	956	996	1037	1078	1119	1159
160	736	778	819	861	902	943	984	1025
180	584	626	668	710	752	793	835	876
200	418	460	503	545	587	629	672	713
220	238	281	324	367	410	453	495	538
240	44	87	131	174	218	261	304	347
260	-163	-119	-75	-30	12	56	100	144
280	-383	-338	-293	-248	-204	-159	-115	-71
300	-615	-570	-524	-479	-434	-388	-344	-299
320	-857	-810	-764	-718	-672	-626	-581	-536
340	-1114	-1066	-1019	-973	-926	-880	-834	-788
360	-1379	-1331	-1283	-1236	-1188	-1141	-1094	-1048
380	-1656	-1607	-1558	-1510	-1462	-1414	-1367	-1319
400	-1941	-1891	-1842	-1793	-1744	-1695	-1647	-1599
420	-2237	-2186	-2136	-2086	-2036	-1987	-1938	-1889
440	-2546	-2494	-2443	-2392	-2342	-2292	-2242	-2192
460	-2861	-2809	-2757	-2705	-2654	-2603	-2552	-2501
480	-3187	-3133	-3080	-3028	-2975	-2923	-2872	-2820
500	-3520	-3466	-3412	-3358	-3305	-3252	-3199	-3147
520	-3864	-3808	-3753	-3699	-3644	-3590	-3537	-3484
540	-4220	-4163	-4107	-4052	-3996	-3941	-3887	-3833
560	-4579	-4521	-4464	-4407	-4351	-4295	-4240	-4185
580	-4949	-4890	-4832	-4774	-4717	-4660	-4603	-4547
600	-5329	-5269	-5209	-5150	-5092	-5034	-4976	-4919
620	-5712	-5651	-5590	-5530	-5470	-5411	-5353	-5295
640	-6103	-6040	-5978	-5917	-5856	-5796	-5736	-5677
660	-6498	-6434	-6371	-6308	-6246	-6185	-6124	-6064
680	-6905	-6840	-6775	-6711	-6648	-6586	-6524	-6463
700	-7315	-7248	-7183	-7117	-7053	-6989	-6926	-6864
720	-7735	-7667	-7600	-7534	-7468	-7403	-7339	-7275
740	-8160	-8091	-8022	-7954	-7887	-7821	-7756	-7691
760	-8585	-8514	-8444	-8375	-8307	-8239	-8173	-8107
780	-9027	-8955	-8883	-8813	-8743	-8674	-8606	-8539
800	-9468	-9394	-9321	-9249	-9178	-9108	-9039	-8970
820	-9918	-9843	-9768	-9695	-9622	-9551	-9480	-9411
840	-10374	-10296	-10220	-10146	-10072	-9999	-9927	-9856
860	-10837	-10758	-10681	-10605	-10529	-10455	-10382	-10310
880	-11306	-11226	-11146	-11069	-10992	-10917	-10842	-10769
900	-11781	-11699	-11618	-11539	-11461	-11384	-11308	-11234
920	-12263	-12179	-12097	-12017	-11937	-11859	-11782	-11706
940	-12749	-12664	-12581	-12499	-12418	-12338	-12260	-12183
960	-13242	-13156	-13071	-12988	-12906	-12825	-12746	-12667
980	-13739	-13651	-13565	-13480	-13397	-13315	-13235	-13156
1000	-14240	-14151	-14064	-13978	-13894	-13811	-13730	-13650

TABLE 2
GIBBS FREE ENERGY IN CALORIES PER MOLE

	PRESSURE IN BARS							
TEMP	4100	4200	4300	4400	4500	4600	4700	4800
20	1637	1675	1713	1751	1789	1827	1864	1902
40	1614	1653	1691	1729	1767	1806	1844	1882
60	1568	1607	1645	1684	1723	1761	1800	1838
80	1503	1542	1581	1620	1659	1698	1737	1776
100	1419	1459	1498	1537	1577	1616	1655	1694
120	1316	1356	1396	1436	1476	1515	1555	1594
140	1200	1240	1280	1320	1360	1401	1440	1480
160	1066	1107	1148	1188	1229	1269	1309	1350
180	917	959	1000	1041	1082	1123	1164	1204
200	755	797	839	880	922	963	1004	1045
220	580	622	665	707	748	790	832	874
240	390	433	475	518	560	603	645	687
260	187	230	273	316	359	402	445	487
280	-27	16	60	103	147	190	233	276
300	-254	-210	-166	-121	-77	-34	9	53
320	-490	-445	-400	-356	-311	-267	-222	-178
340	-742	-696	-651	-605	-560	-515	-470	-425
360	-1001	-955	-909	-863	-817	-771	-726	-680
380	-1272	-1225	-1178	-1131	-1085	-1039	-993	-947
400	-1551	-1503	-1455	-1408	-1361	-1314	-1267	-1221
420	-1840	-1792	-1743	-1695	-1648	-1600	-1553	-1505
440	-2143	-2093	-2044	-1996	-1947	-1899	-1851	-1803
460	-2451	-2401	-2351	-2302	-2252	-2203	-2155	-2106
480	-2769	-2718	-2668	-2618	-2568	-2518	-2468	-2419
500	-3095	-3043	-2992	-2941	-2890	-2839	-2789	-2739
520	-3431	-3378	-3326	-3274	-3223	-3171	-3120	-3069
540	-3779	-3726	-3673	-3620	-3567	-3515	-3463	-3412
560	-4130	-4076	-4022	-3968	-3915	-3862	-3809	-3756
580	-4492	-4436	-4382	-4327	-4273	-4219	-4165	-4112
600	-4863	-4806	-4750	-4695	-4640	-4585	-4531	-4477
620	-5237	-5180	-5123	-5067	-5011	-4955	-4900	-4845
640	-5618	-5560	-5502	-5445	-5388	-5331	-5275	-5219
660	-6004	-5945	-5886	-5828	-5770	-5713	-5655	-5599
680	-6402	-6342	-6282	-6222	-6164	-6105	-6047	-5990
700	-6802	-6741	-6680	-6620	-6560	-6500	-6442	-6383
720	-7212	-7150	-7088	-7027	-6966	-6905	-6846	-6786
740	-7627	-7563	-7500	-7438	-7376	-7315	-7254	-7194
760	-8041	-7977	-7913	-7849	-7787	-7724	-7662	-7601
780	-8473	-8407	-8342	-8277	-8213	-8150	-8087	-8025
800	-8903	-8836	-8769	-8704	-8639	-8574	-8510	-8447
820	-9342	-9274	-9206	-9139	-9073	-9008	-8943	-8879
840	-9786	-9717	-9648	-9580	-9513	-9447	-9381	-9315
860	-10238	-10168	-10098	-10029	-9961	-9893	-9827	-9760
880	-10696	-10624	-10553	-10483	-10414	-10345	-10277	-10210
900	-11160	-11087	-11015	-10944	-10873	-10804	-10735	-10667
920	-11631	-11557	-11484	-11412	-11340	-11270	-11200	-11131
940	-12107	-12032	-11958	-11885	-11812	-11741	-11670	-11600
960	-12590	-12514	-12439	-12365	-12292	-12219	-12147	-12076
980	-13078	-13001	-12925	-12850	-12776	-12702	-12630	-12558
1000	-13571	-13493	-13416	-13341	-13266	-13192	-13119	-13047

TABLE 2
GIBBS FREE ENERGY IN CALORIES PER MOLE

	PRESSURE IN BARS							
TEMP	4900	5000	5100	5200	5300	5400	5500	5600
20	1940	1977	2015	2052	2089	2126	2164	2201
40	1920	1957	1995	2033	2071	2108	2146	2183
60	1876	1914	1953	1991	2029	2066	2104	2142
80	1814	1853	1891	1930	1968	2006	2045	2083
100	1733	1772	1811	1850	1889	1927	1966	2004
120	1634	1673	1712	1751	1790	1829	1868	1907
140	1520	1560	1599	1639	1678	1718	1757	1796
160	1390	1430	1470	1510	1550	1590	1629	1669
180	1245	1285	1326	1366	1407	1447	1487	1527
200	1086	1127	1168	1209	1250	1290	1331	1371
220	915	957	998	1039	1080	1121	1162	1203
240	729	771	812	854	896	937	979	1020
260	530	572	615	657	699	741	783	824
280	319	362	405	448	490	533	575	617
300	96	140	183	226	269	312	355	398
320	-134	-90	-46	-3	40	83	127	170
340	-381	-336	-292	-248	-203	-159	-116	-72
360	-635	-590	-545	-500	-456	-411	-367	-322
380	-901	-855	-810	-764	-719	-674	-629	-584
400	-1174	-1128	-1082	-1036	-990	-944	-899	-853
420	-1458	-1411	-1364	-1318	-1271	-1225	-1179	-1133
440	-1755	-1707	-1660	-1613	-1566	-1519	-1472	-1426
460	-2057	-2009	-1961	-1913	-1866	-1818	-1771	-1724
480	-2370	-2321	-2272	-2223	-2175	-2127	-2079	-2031
500	-2689	-2639	-2590	-2541	-2492	-2443	-2394	-2346
520	-3019	-2968	-2918	-2868	-2818	-2769	-2720	-2670
540	-3360	-3309	-3258	-3208	-3157	-3107	-3057	-3007
560	-3704	-3652	-3601	-3549	-3498	-3447	-3397	-3346
580	-4059	-4006	-3954	-3902	-3850	-3798	-3747	-3696
600	-4423	-4369	-4316	-4263	-4211	-4158	-4106	-4054
620	-4790	-4736	-4682	-4628	-4575	-4522	-4469	-4416
640	-5164	-5109	-5054	-5000	-4945	-4892	-4838	-4785
660	-5542	-5486	-5431	-5376	-5321	-5266	-5212	-5157
680	-5933	-5876	-5819	-5763	-5707	-5652	-5597	-5542
700	-6325	-6267	-6210	-6153	-6096	-6040	-5984	-5929
720	-6727	-6669	-6610	-6552	-6495	-6438	-6381	-6325
740	-7134	-7074	-7015	-6957	-6898	-6840	-6783	-6726
760	-7540	-7480	-7420	-7360	-7301	-7242	-7184	-7126
780	-7963	-7902	-7841	-7780	-7720	-7661	-7601	-7543
800	-8384	-8322	-8260	-8199	-8138	-8077	-8017	-7958
820	-8815	-8752	-8689	-8627	-8565	-8504	-8443	-8382
840	-9251	-9187	-9123	-9060	-8997	-8935	-8873	-8812
860	-9694	-9629	-9565	-9501	-9437	-9374	-9311	-9249
880	-10143	-10077	-10012	-9947	-9882	-9818	-9755	-9692
900	-10599	-10532	-10465	-10400	-10334	-10269	-10205	-10141
920	-11062	-10994	-10927	-10860	-10794	-10728	-10663	-10598
940	-11530	-11461	-11393	-11326	-11259	-11192	-11126	-11061
960	-12006	-11937	-11868	-11799	-11732	-11665	-11598	-11532
980	-12487	-12417	-12347	-12279	-12210	-12142	-12075	-12008
1000	-12975	-12904	-12834	-12765	-12696	-12627	-12560	-12492

TABLE 2
GIBBS FREE ENERGY IN CALORIES PER MOLE

	PRESSURE IN BARS							
TEMP	5700	5800	5900	6000	6100	6200	6300	6400
20	2238	2275	2312	2348	2385	2422	2459	2495
40	2220	2258	2295	2332	2369	2406	2443	2480
60	2180	2217	2255	2293	2330	2367	2405	2442
80	2121	2159	2197	2235	2272	2310	2348	2385
100	2043	2081	2119	2157	2196	2234	2272	2310
120	1946	1985	2023	2062	2100	2139	2177	2215
140	1836	1875	1914	1953	1991	2030	2069	2108
160	1709	1748	1787	1827	1866	1905	1944	1983
180	1567	1607	1646	1686	1726	1765	1805	1844
200	1412	1452	1492	1532	1572	1612	1652	1692
220	1244	1285	1325	1366	1406	1447	1487	1527
240	1061	1102	1143	1184	1225	1266	1307	1347
260	866	908	949	991	1032	1073	1114	1155
280	659	701	743	785	827	869	910	952
300	440	483	525	568	610	652	694	736
320	213	256	299	342	384	427	470	512
340	-28	14	58	101	144	187	230	273
360	-278	-234	-190	-147	-103	-59	-16	26
380	-539	-495	-450	-406	-362	-318	-274	-230
400	-808	-763	-718	-673	-629	-584	-540	-495
420	-1087	-1042	-996	-951	-906	-860	-815	-771
440	-1379	-1333	-1287	-1241	-1195	-1149	-1104	-1058
460	-1677	-1630	-1583	-1537	-1490	-1444	-1398	-1352
480	-1983	-1936	-1889	-1842	-1795	-1748	-1701	-1655
500	-2298	-2250	-2202	-2154	-2106	-2059	-2012	-1965
520	-2622	-2573	-2524	-2476	-2428	-2380	-2332	-2284
540	-2958	-2908	-2859	-2810	-2761	-2713	-2664	-2616
560	-3296	-3246	-3196	-3146	-3097	-3048	-2999	-2950
580	-3645	-3594	-3544	-3493	-3443	-3393	-3344	-3294
600	-4003	-3951	-3900	-3849	-3798	-3748	-3698	-3648
620	-4364	-4312	-4260	-4208	-4157	-4106	-4055	-4004
640	-4732	-4679	-4626	-4574	-4522	-4470	-4418	-4367
660	-5104	-5050	-4997	-4944	-4891	-4839	-4786	-4734
680	-5487	-5433	-5379	-5325	-5272	-5219	-5166	-5113
700	-5873	-5818	-5763	-5709	-5655	-5601	-5547	-5494
720	-6269	-6213	-6157	-6102	-6047	-5993	-5938	-5884
740	-6669	-6612	-6556	-6500	-6444	-6389	-6334	-6279
760	-7068	-7011	-6954	-6897	-6841	-6785	-6729	-6674
780	-7484	-7426	-7368	-7311	-7254	-7197	-7140	-7084
800	-7898	-7840	-7781	-7723	-7665	-7607	-7550	-7493
820	-8322	-8262	-8203	-8144	-8085	-8027	-7969	-7911
840	-8751	-8690	-8630	-8570	-8511	-8452	-8393	-8335
860	-9187	-9126	-9065	-9004	-8944	-8884	-8825	-8766
880	-9629	-9567	-9505	-9444	-9383	-9323	-9262	-9203
900	-10078	-10015	-9952	-9890	-9829	-9767	-9706	-9646
920	-10534	-10471	-10407	-10344	-10282	-10220	-10158	-10097
940	-10996	-10932	-10868	-10804	-10741	-10678	-10616	-10554
960	-11466	-11401	-11336	-11272	-11208	-11145	-11082	-11019
980	-11942	-11876	-11811	-11746	-11682	-11618	-11554	-11491
1000	-12425	-12359	-12293	-12228	-12163	-12098	-12034	-11970

TABLE 2
GIBBS FREE ENERGY IN CALORIES PER MOLE

TEMP	PRESSURE IN BARS							
	6500	6600	6700	6800	6900	7000	7100	7200
20	2532	2568	2605	2641	2678	2714	2750	2786
40	2517	2554	2591	2628	2664	2701	2737	2774
60	2479	2516	2553	2590	2627	2664	2701	2738
80	2423	2460	2498	2535	2572	2610	2647	2684
100	2347	2385	2423	2461	2498	2536	2574	2611
120	2254	2292	2330	2368	2406	2444	2482	2520
140	2146	2185	2223	2262	2300	2338	2377	2415
160	2022	2061	2100	2139	2178	2216	2255	2293
180	1884	1923	1962	2001	2040	2079	2118	2157
200	1732	1771	1811	1851	1890	1929	1969	2008
220	1567	1607	1647	1687	1727	1767	1807	1846
240	1388	1428	1469	1509	1549	1590	1630	1670
260	1196	1237	1278	1319	1359	1400	1440	1481
280	993	1034	1076	1117	1158	1199	1240	1281
300	778	820	861	903	944	986	1027	1068
320	554	596	639	681	722	764	806	848
340	316	358	401	443	486	528	570	612
360	70	113	156	199	242	284	327	369
380	-186	-143	-99	-56	-13	30	73	116
400	-451	-407	-363	-319	-275	-232	-188	-145
420	-726	-681	-637	-592	-548	-504	-460	-416
440	-1013	-968	-923	-878	-833	-789	-744	-700
460	-1306	-1260	-1215	-1170	-1124	-1079	-1034	-989
480	-1608	-1562	-1516	-1470	-1424	-1379	-1333	-1288
500	-1918	-1871	-1824	-1778	-1732	-1685	-1639	-1593
520	-2237	-2190	-2142	-2095	-2048	-2002	-1955	-1909
540	-2568	-2520	-2472	-2425	-2377	-2330	-2283	-2236
560	-2901	-2853	-2804	-2756	-2708	-2660	-2613	-2565
580	-3245	-3196	-3147	-3098	-3050	-3001	-2953	-2905
600	-3598	-3548	-3498	-3449	-3400	-3351	-3302	-3254
620	-3954	-3903	-3853	-3803	-3753	-3704	-3655	-3605
640	-4316	-4265	-4214	-4164	-4113	-4063	-4013	-3963
660	-4682	-4631	-4579	-4528	-4477	-4427	-4376	-4326
680	-5061	-5008	-4956	-4905	-4853	-4802	-4751	-4700
700	-5441	-5388	-5335	-5283	-5231	-5179	-5127	-5076
720	-5831	-5777	-5724	-5671	-5618	-5565	-5513	-5461
740	-6225	-6171	-6117	-6063	-6010	-5956	-5903	-5851
760	-6618	-6564	-6509	-6455	-6401	-6347	-6293	-6240
780	-7028	-6973	-6918	-6863	-6808	-6753	-6699	-6645
800	-7437	-7380	-7324	-7269	-7213	-7158	-7103	-7049
820	-7854	-7797	-7740	-7684	-7628	-7572	-7517	-7462
840	-8277	-8219	-8162	-8105	-8048	-7991	-7935	-7879
860	-8707	-8649	-8591	-8533	-8476	-8419	-8362	-8305
880	-9143	-9084	-9025	-8967	-8909	-8851	-8793	-8736
900	-9586	-9526	-9466	-9407	-9348	-9290	-9232	-9174
920	-10036	-9976	-9916	-9856	-9796	-9737	-9678	-9620
940	-10492	-10431	-10370	-10310	-10250	-10190	-10130	-10071
960	-10957	-10895	-10834	-10772	-10712	-10651	-10591	-10531
980	-11428	-11365	-11303	-11241	-11179	-11118	-11057	-10997
1000	-11906	-11843	-11780	-11718	-11655	-11593	-11532	-11470

TABLE 2
GIBBS FREE ENERGY IN CALORIES PER MOLE

TEMP	PRESSURE IN BARS 7300	7400	7500	7600	7700	7800	7900	8000
20	2822	2859	2895	2931	2966	3002	3038	3074
40	2810	2847	2883	2919	2956	2992	3028	3064
60	2775	2812	2848	2885	2921	2958	2994	3031
80	2721	2758	2795	2832	2869	2906	2943	2979
100	2648	2686	2723	2760	2798	2835	2872	2909
120	2557	2595	2633	2670	2708	2745	2783	2820
140	2453	2491	2529	2567	2605	2642	2680	2718
160	2332	2370	2409	2447	2485	2523	2561	2599
180	2196	2235	2274	2312	2351	2389	2428	2466
200	2047	2086	2125	2164	2203	2242	2281	2320
220	1886	1925	1965	2004	2043	2083	2122	2161
240	1710	1750	1789	1829	1869	1908	1948	1987
260	1521	1562	1602	1642	1682	1722	1762	1802
280	1321	1362	1403	1443	1484	1524	1564	1605
300	1109	1151	1192	1232	1273	1314	1355	1395
320	889	931	972	1014	1055	1096	1137	1178
340	654	696	738	780	821	863	905	946
360	412	454	496	539	581	623	665	706
380	159	201	244	287	329	372	414	456
400	-101	-58	-15	27	70	113	156	198
420	-372	-328	-285	-241	-198	-155	-112	-68
440	-656	-611	-567	-523	-480	-436	-392	-349
460	-944	-900	-855	-811	-766	-722	-678	-634
480	-1242	-1197	-1152	-1107	-1063	-1018	-973	-929
500	-1548	-1502	-1456	-1411	-1366	-1321	-1276	-1231
520	-1862	-1816	-1770	-1724	-1679	-1633	-1587	-1542
540	-2189	-2142	-2096	-2050	-2003	-1957	-1911	-1865
560	-2518	-2471	-2424	-2377	-2330	-2283	-2237	-2191
580	-2857	-2809	-2762	-2714	-2667	-2620	-2573	-2526
600	-3205	-3157	-3109	-3061	-3013	-2965	-2918	-2871
620	-3556	-3508	-3459	-3410	-3362	-3314	-3266	-3218
640	-3914	-3864	-3815	-3766	-3717	-3669	-3620	-3572
660	-4276	-4226	-4176	-4126	-4077	-4028	-3979	-3930
680	-4649	-4598	-4548	-4498	-4448	-4398	-4349	-4299
700	-5024	-4973	-4922	-4872	-4821	-4771	-4721	-4671
720	-5409	-5357	-5306	-5255	-5204	-5153	-5102	-5052
740	-5798	-5746	-5694	-5642	-5590	-5539	-5488	-5437
760	-6187	-6134	-6081	-6029	-5977	-5925	-5873	-5821
780	-6591	-6538	-6485	-6432	-6379	-6327	-6274	-6222
800	-6994	-6940	-6887	-6833	-6780	-6727	-6674	-6621
820	-7407	-7352	-7297	-7243	-7189	-7136	-7082	-7029
840	-7824	-7769	-7713	-7659	-7604	-7550	-7496	-7442
860	-8249	-8193	-8137	-8082	-8027	-7972	-7917	-7863
880	-8679	-8623	-8566	-8510	-8455	-8399	-8344	-8289
900	-9116	-9059	-9002	-8946	-8889	-8833	-8777	-8722
920	-9562	-9504	-9446	-9389	-9332	-9275	-9219	-9162
940	-10012	-9954	-9895	-9837	-9780	-9722	-9665	-9608
960	-10471	-10412	-10353	-10294	-10236	-10178	-10120	-10063
980	-10936	-10876	-10817	-10757	-10698	-10639	-10581	-10522
1000	-11409	-11349	-11288	-11228	-11168	-11108	-11049	-10990

TABLE 2
GIBBS FREE ENERGY IN CALORIES PER MOLE

TEMP	PRESSURE IN BARS 8100	8200	8300	8400	8500	8600	8700	8800
20	3110	3145	3181	3217	3252	3288	3323	3358
40	3100	3136	3172	3208	3244	3280	3315	3351
60	3067	3104	3140	3176	3212	3248	3284	3320
80	3016	3053	3089	3126	3162	3199	3235	3271
100	2946	2983	3020	3057	3093	3130	3167	3204
120	2857	2895	2932	2969	3006	3043	3080	3117
140	2756	2793	2831	2868	2906	2943	2980	3018
160	2637	2675	2713	2751	2789	2827	2864	2902
180	2505	2543	2581	2619	2657	2695	2733	2771
200	2358	2397	2436	2474	2513	2551	2589	2628
220	2200	2239	2278	2317	2356	2394	2433	2472
240	2027	2066	2106	2145	2184	2223	2262	2301
260	1841	1881	1921	1960	2000	2039	2079	2118
280	1645	1685	1725	1765	1805	1844	1884	1924
300	1436	1476	1517	1557	1597	1638	1678	1718
320	1219	1260	1301	1341	1382	1423	1463	1503
340	987	1029	1070	1111	1152	1193	1234	1274
360	748	790	831	873	914	956	997	1038
380	498	540	582	624	666	708	749	791
400	241	283	326	368	410	452	494	536
420	-25	16	59	102	145	187	230	272
440	-305	-262	-219	-175	-132	-89	-47	-4
460	-590	-547	-503	-459	-416	-373	-329	-286
480	-885	-840	-796	-752	-708	-665	-621	-577
500	-1186	-1141	-1097	-1052	-1008	-964	-920	-876
520	-1497	-1452	-1407	-1362	-1317	-1272	-1228	-1184
540	-1820	-1774	-1729	-1683	-1638	-1593	-1548	-1503
560	-2144	-2098	-2052	-2007	-1961	-1915	-1870	-1825
580	-2480	-2433	-2387	-2340	-2294	-2248	-2202	-2157
600	-2823	-2776	-2730	-2683	-2636	-2590	-2543	-2497
620	-3171	-3123	-3076	-3028	-2981	-2934	-2888	-2841
640	-3524	-3476	-3428	-3380	-3333	-3285	-3238	-3191
660	-3881	-3833	-3784	-3736	-3688	-3640	-3593	-3545
680	-4250	-4201	-4152	-4104	-4055	-4007	-3959	-3911
700	-4621	-4572	-4522	-4473	-4424	-4375	-4327	-4278
720	-5002	-4951	-4902	-4852	-4802	-4753	-4704	-4655
740	-5386	-5336	-5285	-5235	-5185	-5135	-5086	-5036
760	-5770	-5719	-5668	-5617	-5567	-5517	-5466	-5416
780	-6170	-6119	-6067	-6016	-5965	-5914	-5864	-5813
800	-6569	-6517	-6465	-6413	-6361	-6310	-6259	-6208
820	-6976	-6923	-6871	-6819	-6767	-6715	-6663	-6612
840	-7389	-7335	-7282	-7229	-7177	-7124	-7072	-7020
860	-7809	-7755	-7701	-7648	-7595	-7542	-7489	-7437
880	-8234	-8180	-8126	-8072	-8018	-7965	-7911	-7858
900	-8666	-8611	-8557	-8502	-8448	-8394	-8340	-8287
920	-9106	-9051	-8995	-8940	-8885	-8831	-8776	-8722
940	-9552	-9495	-9439	-9384	-9328	-9273	-9218	-9163
960	-10005	-9948	-9892	-9835	-9779	-9723	-9667	-9612
980	-10464	-10407	-10349	-10292	-10235	-10178	-10122	-10066
1000	-10931	-10872	-10814	-10756	-10698	-10641	-10583	-10526

TABLE 2
GIBBS FREE ENERGY IN CALORIES PER MOLE

	PRESSURE IN BARS							
TEMP	8900	9000	9100	9200	9300	9400	9500	9600
20	(3394)	(3429)	(3464)	(3499)	(3535)	(3570)	(3605)	(3640)
40	3387	3422	3458	3494	3529	3564	3600	3635
60	3356	3392	3428	3464	3500	3536	3571	3607
80	3308	3344	3380	3416	3452	3489	3525	3561
100	3240	3277	3313	3350	3386	3423	3459	3495
120	3154	3191	3228	3265	3301	3338	3375	3411
140	3055	3092	3129	3166	3203	3240	3277	3314
160	2939	2977	3014	3052	3089	3126	3164	3201
180	2809	2847	2885	2923	2960	2998	3035	3073
200	2666	2704	2742	2780	2818	2856	2894	2932
220	2510	2549	2587	2626	2664	2702	2740	2779
240	2340	2379	2418	2456	2495	2534	2572	2611
260	2157	2197	2236	2275	2314	2353	2392	2430
280	1964	2003	2043	2082	2121	2161	2200	2239
300	1758	1798	1837	1877	1917	1957	1996	2036
320	1544	1584	1624	1664	1704	1744	1784	1824
340	1315	1356	1396	1437	1477	1518	1558	1598
360	1079	1120	1161	1202	1243	1283	1324	1365
380	832	874	915	956	998	1039	1080	1121
400	578	620	662	703	745	786	828	869
420	314	356	398	440	482	524	566	607
440	38	80	123	165	207	250	292	334
460	-243	-200	-157	-115	-72	-29	12	55
480	-534	-491	-447	-404	-361	-318	-275	-233
500	-832	-788	-744	-701	-657	-614	-571	-528
520	-1139	-1095	-1051	-1007	-963	-919	-876	-832
540	-1458	-1414	-1369	-1325	-1281	-1237	-1193	-1149
560	-1780	-1735	-1690	-1645	-1600	-1556	-1511	-1467
580	-2111	-2066	-2020	-1975	-1930	-1885	-1840	-1796
600	-2451	-2405	-2360	-2314	-2269	-2223	-2178	-2133
620	-2795	-2748	-2702	-2656	-2610	-2564	-2519	-2473
640	-3144	-3097	-3051	-3004	-2958	-2912	-2866	-2820
660	-3498	-3450	-3403	-3356	-3310	-3263	-3217	-3170
680	-3863	-3815	-3767	-3720	-3673	-3626	-3579	-3532
700	-4230	-4182	-4134	-4086	-4038	-3991	-3943	-3896
720	-4606	-4558	-4509	-4461	-4413	-4365	-4317	-4269
740	-4987	-4938	-4889	-4840	-4791	-4743	-4695	-4647
760	-5367	-5317	-5268	-5219	-5170	-5121	-5072	-5023
780	-5763	-5713	-5663	-5613	-5564	-5514	-5465	-5416
800	-6157	-6107	-6056	-6006	-5956	-5906	-5857	-5807
820	-6560	-6509	-6459	-6408	-6357	-6307	-6257	-6207
840	-6969	-6917	-6866	-6815	-6764	-6713	-6663	-6612
860	-7385	-7333	-7281	-7229	-7178	-7127	-7076	-7025
880	-7806	-7753	-7701	-7649	-7597	-7545	-7494	-7442
900	-8233	-8180	-8127	-8075	-8022	-7970	-7918	-7866
920	-8668	-8615	-8561	-8508	-8455	-8402	-8350	-8298
940	-9108	-9054	-9000	-8946	-8893	-8840	-8786	-8734
960	-9557	-9502	-9447	-9392	-9338	-9284	-9231	-9177
980	-10010	-9954	-9898	-9843	-9788	-9734	-9679	-9625
1000	-10469	-10413	-10356	-10300	-10245	-10189	-10134	-10079

TABLE 2
GIBBS FREE ENERGY IN CALORIES PER MOLE

TEMP	PRESSURE IN BARS 9700	9800	9900	10000
20	(3675)	(3709)	(3744)	(3779)
40	3670	3706	3741	3776
60	3643	3678	3714	3749
80	3597	3632	3668	3704
100	3531	3568	3604	3640
120	3448	3484	3521	3557
140	3351	3388	3425	3461
160	3238	3275	3312	3349
180	3110	3148	3185	3222
200	2970	3007	3045	3083
220	2817	2855	2893	2931
240	2649	2687	2726	2764
260	2469	2508	2547	2585
280	2278	2317	2356	2395
300	2075	2114	2154	2193
320	1864	1904	1943	1983
340	1638	1678	1718	1758
360	1405	1446	1486	1526
380	1161	1202	1243	1283
400	910	951	992	1033
420	649	690	732	773
440	376	417	459	501
460	97	139	181	223
480	-190	-147	-105	-63
500	-485	-442	-399	-356
520	-789	-746	-703	-659
540	-1105	-1061	-1018	-974
560	-1423	-1379	-1335	-1291
580	-1751	-1707	-1662	-1618
600	-2088	-2043	-1998	-1954
620	-2428	-2383	-2338	-2293
640	-2774	-2728	-2683	-2637
660	-3124	-3078	-3032	-2986
680	-3486	-3439	-3393	-3347
700	-3849	-3802	-3755	-3709
720	-4222	-4175	-4127	-4080
740	-4599	-4551	-4504	-4456
760	-4975	-4927	-4879	-4831
780	-5368	-5319	-5271	-5222
800	-5758	-5709	-5660	-5612
820	-6158	-6108	-6059	-6010
840	-6562	-6512	-6462	-6413
860	-6974	-6924	-6874	-6824
880	-7391	-7341	-7290	-7240
900	-7815	-7763	-7712	-7661
920	-8246	-8194	-8142	-8091
940	-8681	-8629	-8576	-8525
960	-9124	-9071	-9018	-8966
980	-9571	-9517	-9464	-9411
1000	-10024	-9969	-9915	-9861

TABLE 3
ENTROPY IN CALORIES PER MOLE PER DEGREE KELVIN

TEMP	PRESSURE IN BARS 100	200	300	400	500	600	700	800
20	1.265	1.256	1.248	1.235	1.222	1.213	1.201	1.188
40	2.444	2.427	2.410	2.392	2.375	2.358	2.341	2.324
60	3.554	3.533	3.507	3.485	3.464	3.442	3.421	3.399
80	4.600	4.570	4.544	4.518	4.492	4.466	4.441	4.419
100	5.589	5.559	5.529	5.495	5.465	5.439	5.409	5.379
120	6.536	6.497	6.459	6.424	6.390	6.355	6.325	6.291
140	7.440	7.397	7.354	7.315	7.276	7.237	7.199	7.164
160	8.309	8.262	8.214	8.171	8.128	8.085	8.046	8.003
180	9.152	9.096	9.045	8.997	8.946	8.898	8.855	8.812
200	9.974	9.910	9.849	9.793	9.737	9.686	9.634	9.587
220	10.774	10.701	10.632	10.568	10.508	10.447	10.391	10.340
240	11.566	11.480	11.403	11.325	11.256	11.192	11.132	11.071
260	12.354	12.250	12.156	12.070	11.992	11.919	11.850	11.781
280	13.150	13.021	12.909	12.810	12.715	12.633	12.552	12.478
300	13.976	13.808	13.666	13.541	13.429	13.330	13.236	13.154
320	24.582	14.630	14.436	14.286	14.152	14.032	13.920	13.821
340	25.297	15.533	15.254	15.047	14.879	14.733	14.604	14.488
360	25.856	16.687	16.140	15.848	15.628	15.447	15.292	15.155
380	26.321	22.878	17.233	16.712	16.411	16.187	16.007	15.848
400	26.734	23.907	19.286	17.724	17.250	16.949	16.721	16.532
420	27.108	24.608	21.794	19.023	18.167	17.732	17.431	17.203
440	27.457	25.163	22.995	20.585	19.208	18.567	18.171	17.883
460	27.779	25.632	23.795	21.885	20.331	19.466	18.946	18.584
480	28.085	26.045	24.415	22.827	21.381	20.374	19.737	19.298
500	28.382	26.428	24.935	23.554	22.272	21.243	20.520	20.013
520	28.666	26.781	25.387	24.152	23.008	22.022	21.265	20.706
540	28.937	27.108	25.796	24.660	23.623	22.706	21.953	21.364
560	29.199	27.418	26.166	25.107	24.156	23.304	22.573	21.979
580	29.453	27.711	26.506	25.508	24.621	23.829	23.137	22.547
600	29.699	27.990	26.824	25.878	25.047	24.303	23.640	23.068
620	29.935	28.257	27.130	26.222	25.430	24.729	24.100	23.550
640	30.168	28.515	27.418	26.540	25.787	25.120	24.522	23.993
660	30.396	28.765	27.693	26.846	26.123	25.486	24.914	24.402
680	30.619	29.006	27.956	27.134	26.441	25.826	25.279	24.789
700	30.835	29.242	28.214	27.414	26.738	26.149	25.624	25.150
720	31.050	29.470	28.459	27.680	27.026	26.458	25.951	25.490
740	31.261	29.694	28.700	27.939	27.302	26.751	26.256	25.817
760	31.463	29.918	28.937	28.188	27.569	27.031	26.553	26.123
780	31.665	30.129	29.165	28.429	27.822	27.302	26.837	26.420
800	31.867	30.340	29.384	28.666	28.072	27.560	27.108	26.704

TABLE 3
ENTROPY IN CALORIES PER MOLE PER DEGREE KELVIN

TEMP	PRESSURE IN BARS							
	900	1000	1100	1200	1300	1400	1500	1600
20	1.175	1.162	1.140	1.120	1.099	1.079	1.060	1.040
40	2.306	2.289	2.271	2.253	2.235	2.217	2.199	2.181
60	3.382	3.361	3.341	3.322	3.303	3.285	3.266	3.247
80	4.393	4.367	4.345	4.323	4.302	4.280	4.259	4.238
100	5.353	5.323	5.297	5.271	5.246	5.220	5.196	5.171
120	6.261	6.231	6.200	6.171	6.142	6.113	6.085	6.057
140	7.130	7.095	7.061	7.028	6.995	6.963	6.931	6.900
160	7.965	7.930	7.893	7.856	7.820	7.784	7.749	7.715
180	8.769	8.731	8.690	8.650	8.610	8.572	8.535	8.498
200	9.540	9.496	9.452	9.409	9.367	9.326	9.286	9.246
220	10.288	10.237	10.189	10.143	10.098	10.054	10.011	9.969
240	11.015	10.959	10.908	10.859	10.811	10.764	10.718	10.674
260	11.721	11.661	11.605	11.551	11.500	11.450	11.401	11.353
280	12.410	12.341	12.279	12.220	12.164	12.110	12.058	12.007
300	13.072	12.995	12.926	12.861	12.800	12.741	12.685	12.630
320	13.726	13.640	13.563	13.490	13.423	13.358	13.297	13.238
340	14.385	14.286	14.198	14.117	14.041	13.970	13.903	13.839
360	15.034	14.927	14.827	14.735	14.651	14.572	14.498	14.428
380	15.714	15.589	15.475	15.372	15.277	15.190	15.108	15.031
400	16.372	16.235	16.105	15.989	15.883	15.785	15.695	15.610
420	17.014	16.850	16.710	16.585	16.471	16.364	16.264	16.171
440	17.655	17.465	17.304	17.161	17.033	16.915	16.805	16.704
460	18.309	18.089	17.901	17.737	17.593	17.462	17.341	17.230
480	18.971	18.709	18.491	18.305	18.143	17.998	17.866	17.745
500	19.634	19.333	19.085	18.877	18.696	18.537	18.393	18.262
520	20.284	19.948	19.674	19.443	19.244	19.070	18.915	18.774
540	20.908	20.542	20.243	19.992	19.777	19.588	19.421	19.270
560	21.502	21.114	20.795	20.526	20.295	20.094	19.915	19.755
580	22.061	21.656	21.321	21.037	20.793	20.580	20.391	20.221
600	22.582	22.173	21.826	21.531	21.275	21.052	20.853	20.674
620	23.068	22.659	22.306	22.001	21.737	21.505	21.298	21.112
640	23.524	23.115	22.759	22.449	22.178	21.939	21.725	21.532
660	23.950	23.545	23.190	22.877	22.602	22.358	22.139	21.941
680	24.346	23.954	23.602	23.290	23.013	22.765	22.542	22.340
700	24.724	24.337	23.991	23.681	23.403	23.154	22.929	22.724
720	25.077	24.703	24.364	24.057	23.781	23.532	23.305	23.099
740	25.413	25.051	24.720	24.419	24.145	23.897	23.670	23.463
760	25.736	25.378	25.055	24.759	24.489	24.242	24.016	23.809
780	26.041	25.697	25.381	25.090	24.823	24.579	24.355	24.148
800	26.334	25.998	25.688	25.403	25.140	24.898	24.675	24.469
820		26.296	25.992	25.710	25.450	25.211	24.989	24.784
840		26.575	26.275	25.997	25.740	25.502	25.281	25.077
860		26.841	26.544	26.269	26.014	25.778	25.558	25.354
880		27.096	26.803	26.531	26.278	26.042	25.823	25.620
900		27.341	27.071	26.823	26.593	26.379	26.178	25.990
920		27.576	27.320	27.083	26.861	26.652	26.456	26.270
940		27.802	27.557	27.328	27.112	26.907	26.714	26.530
960		28.019	27.781	27.556	27.343	27.140	26.947	26.764
980		28.227	27.990	27.765	27.550	27.345	27.150	26.964
1000		28.428	28.184	27.951	27.729	27.518	27.316	27.125

TABLE 3
ENTROPY IN CALORIES PER MOLE PER DEGREE KELVIN

PRESSURE IN BARS

TEMP	1700	1800	1900	2000	2100	2200	2300	2400
20	1.021	1.001	0.982	0.963	0.944	0.925	0.906	0.887
40	2.163	2.146	2.128	2.110	2.093	2.075	2.058	2.040
60	3.228	3.210	3.191	3.173	3.155	3.136	3.118	3.100
80	4.217	4.196	4.175	4.155	4.134	4.114	4.095	4.075
100	5.147	5.124	5.100	5.077	5.054	5.032	5.009	4.988
120	6.029	6.003	5.976	5.950	5.924	5.899	5.874	5.850
140	6.870	6.840	6.810	6.781	6.753	6.725	6.698	6.671
160	7.682	7.649	7.617	7.585	7.554	7.523	7.494	7.464
180	8.461	8.426	8.391	8.357	8.323	8.291	8.258	8.227
200	9.208	9.170	9.133	9.096	9.060	9.025	8.991	8.957
220	9.928	9.888	9.848	9.810	9.772	9.735	9.698	9.663
240	10.630	10.587	10.546	10.505	10.465	10.426	10.387	10.350
260	11.307	11.262	11.218	11.175	11.133	11.091	11.051	11.012
280	11.958	11.910	11.864	11.818	11.773	11.730	11.688	11.646
300	12.578	12.526	12.477	12.428	12.381	12.336	12.291	12.247
320	13.181	13.126	13.074	13.022	12.972	12.924	12.877	12.831
340	13.777	13.718	13.662	13.607	13.554	13.502	13.452	13.404
360	14.361	14.297	14.236	14.178	14.121	14.066	14.013	13.962
380	14.958	14.889	14.823	14.760	14.699	14.641	14.585	14.530
400	15.531	15.456	15.384	15.316	15.251	15.189	15.129	15.071
420	16.084	16.003	15.925	15.852	15.764	15.697	15.633	15.572
440	16.609	16.520	16.437	16.358	16.269	16.198	16.130	16.064
460	17.127	17.031	16.941	16.856	16.777	16.701	16.628	16.559
480	17.633	17.530	17.433	17.342	17.276	17.195	17.118	17.044
500	18.142	18.030	17.926	17.829	17.777	17.691	17.609	17.531
520	18.644	18.525	18.414	18.311	18.269	18.178	18.091	18.010
540	19.132	19.005	18.887	18.778	18.741	18.645	18.554	18.469
560	19.608	19.473	19.349	19.233	19.196	19.095	19.000	18.910
580	20.066	19.924	19.794	19.672	19.627	19.522	19.422	19.328
600	20.512	20.364	20.227	20.100	20.041	19.931	19.827	19.730
620	20.943	20.788	20.645	20.513	20.435	20.320	20.213	20.112
640	21.357	21.197	21.050	20.913	20.810	20.692	20.581	20.477
660	21.761	21.596	21.444	21.303	21.173	21.051	20.937	20.830
680	22.156	21.986	21.831	21.687	21.528	21.403	21.286	21.176
700	22.536	22.364	22.205	22.058	21.873	21.745	21.626	21.513
720	22.909	22.735	22.574	22.424	22.216	22.086	21.964	21.849
740	23.272	23.097	22.934	22.782	22.557	22.425	22.301	22.184
760	23.618	23.442	23.278	23.125	22.892	22.758	22.633	22.514
780	23.957	23.780	23.616	23.463	23.232	23.098	22.971	22.851
800	24.278	24.102	23.937	23.783	23.568	23.433	23.304	23.183
820	24.594	24.418	24.253	24.099	23.912	23.776	23.647	23.525
840	24.888	24.711	24.546	24.392	24.245	24.109	23.979	23.856
860	25.165	24.989	24.823	24.670	24.571	24.434	24.305	24.182
880	25.431	25.255	25.090	24.935	24.887	24.751	24.621	24.498
900	25.812	25.644	25.485	25.335	25.192	25.056	24.926	24.803
920	26.095	25.928	25.771	25.620	25.478	25.342	25.213	25.090
940	26.356	26.190	26.033	25.883	25.741	25.606	25.477	25.354
960	26.589	26.423	26.266	26.116	25.973	25.838	25.710	25.588
980	26.787	26.620	26.460	26.309	26.166	26.031	25.903	25.782
1000	26.944	26.772	26.609	26.456	26.311	26.175	26.047	25.927

TABLE 3
ENTROPY IN CALORIES PER MOLE PER DEGREE KELVIN

TEMP	PRESSURE IN BARS 2500	2600	2700	2800	2900	3000	3100	3200
20	0.869	0.850	0.831	0.812	0.794	0.775	0.756	0.738
40	2.023	2.005	1.988	1.970	1.953	1.935	1.918	1.900
60	3.082	3.064	3.046	3.029	3.011	2.993	2.975	2.958
80	4.055	4.036	4.017	3.998	3.979	3.960	3.941	3.922
100	4.966	4.944	4.923	4.902	4.882	4.861	4.841	4.821
120	5.826	5.802	5.779	5.756	5.733	5.711	5.689	5.667
140	6.644	6.618	6.593	6.568	6.543	6.519	6.495	6.471
160	7.436	7.407	7.380	7.353	7.326	7.300	7.274	7.249
180	8.196	8.166	8.136	8.107	8.078	8.050	8.023	7.996
200	8.925	8.892	8.861	8.830	8.799	8.770	8.741	8.712
220	9.628	9.594	9.561	9.528	9.496	9.464	9.434	9.404
240	10.313	10.277	10.242	10.208	10.174	10.141	10.109	10.077
260	10.973	10.935	10.898	10.862	10.827	10.792	10.759	10.725
280	11.606	11.566	11.527	11.489	11.452	11.416	11.381	11.346
300	12.205	12.163	12.123	12.083	12.044	12.006	11.969	11.933
320	12.786	12.742	12.699	12.658	12.617	12.577	12.539	12.501
340	13.356	13.310	13.266	13.222	13.179	13.138	13.097	13.057
360	13.912	13.863	13.816	13.770	13.725	13.682	13.639	13.598
380	14.477	14.426	14.376	14.328	14.281	14.235	14.191	14.147
400	15.015	14.961	14.908	14.858	14.808	14.760	14.714	14.668
420	15.513	15.455	15.400	15.347	15.295	15.245	15.196	15.149
440	16.002	15.941	15.883	15.827	15.773	15.720	15.669	15.619
460	16.493	16.429	16.368	16.309	16.252	16.196	16.143	16.091
480	16.974	16.907	16.843	16.781	16.721	16.664	16.608	16.554
500	17.458	17.387	17.319	17.255	17.192	17.132	17.074	17.018
520	17.932	17.858	17.787	17.719	17.654	17.592	17.531	17.473
540	18.387	18.310	18.236	18.165	18.097	18.032	17.969	17.909
560	18.825	18.744	18.667	18.594	18.523	18.456	18.391	18.328
580	19.240	19.156	19.076	18.999	18.926	18.856	18.789	18.725
600	19.638	19.550	19.468	19.389	19.313	19.241	19.172	19.105
620	20.017	19.926	19.841	19.759	19.681	19.607	19.535	19.467
640	20.378	20.285	20.197	20.113	20.033	19.956	19.883	19.812
660	20.728	20.633	20.542	20.455	20.373	20.294	20.219	20.147
680	21.072	20.974	20.881	20.792	20.708	20.627	20.550	20.476
700	21.407	21.306	21.211	21.120	21.034	20.951	20.872	20.797
720	21.741	21.638	21.541	21.448	21.360	21.275	21.195	21.117
740	22.074	21.969	21.870	21.776	21.686	21.599	21.517	21.438
760	22.402	22.296	22.195	22.099	22.007	21.919	21.835	21.754
780	22.737	22.629	22.526	22.429	22.335	22.246	22.160	22.078
800	23.068	22.959	22.855	22.756	22.661	22.571	22.484	22.400
820	23.409	23.298	23.193	23.093	22.997	22.905	22.817	22.733
840	23.740	23.628	23.522	23.421	23.324	23.231	23.142	23.057
860	24.064	23.952	23.846	23.744	23.646	23.553	23.463	23.377
880	24.381	24.268	24.161	24.059	23.961	23.868	23.778	23.692
900	24.685	24.573	24.466	24.364	24.266	24.173	24.084	23.999
920	24.973	24.861	24.755	24.653	24.556	24.464	24.376	24.292
940	25.238	25.127	25.022	24.921	24.826	24.735	24.649	24.567
960	25.472	25.363	25.259	25.161	25.068	24.980	24.896	24.817
980	25.668	25.560	25.458	25.363	25.273	25.188	25.108	25.033
1000	25.814	25.709	25.610	25.518	25.432	25.352	25.277	25.207

TABLE 3
ENTROPY IN CALORIES PER MOLE PER DEGREE KELVIN

TEMP	PRESSURE IN BARS 3300	3400	3500	3600	3700	3800	3900	4000
20	0.719	0.701	0.682	0.663	0.645	0.626	0.607	0.589
40	1.882	1.865	1.847	1.830	1.812	1.794	1.777	1.759
60	2.940	2.922	2.904	2.887	2.869	2.851	2.834	2.816
80	3.904	3.885	3.867	3.848	3.830	3.811	3.793	3.775
100	4.801	4.781	4.761	4.742	4.722	4.703	4.684	4.664
120	5.646	5.624	5.603	5.582	5.561	5.541	5.520	5.500
140	6.448	6.425	6.403	6.380	6.358	6.336	6.314	6.293
160	7.224	7.199	7.175	7.151	7.128	7.105	7.082	7.059
180	7.969	7.943	7.918	7.893	7.868	7.843	7.819	7.795
200	8.684	8.656	8.629	8.603	8.576	8.551	8.525	8.500
220	9.374	9.345	9.316	9.289	9.261	9.234	9.207	9.181
240	10.046	10.015	9.986	9.956	9.928	9.899	9.871	9.844
260	10.693	10.661	10.630	10.599	10.569	10.539	10.510	10.482
280	11.312	11.279	11.246	11.214	11.183	11.152	11.122	11.092
300	11.897	11.862	11.828	11.795	11.762	11.730	11.699	11.668
320	12.464	12.427	12.392	12.357	12.323	12.289	12.256	12.224
340	13.019	12.981	12.944	12.907	12.872	12.837	12.803	12.769
360	13.557	13.518	13.479	13.441	13.404	13.368	13.332	13.297
380	14.105	14.064	14.023	13.984	13.945	13.908	13.871	13.834
400	14.624	14.581	14.539	14.498	14.458	14.418	14.380	14.342
420	15.102	15.057	15.013	14.971	14.929	14.888	14.848	14.809
440	15.571	15.524	15.479	15.434	15.391	15.348	15.307	15.267
460	16.041	15.992	15.945	15.899	15.854	15.810	15.767	15.725
480	16.502	16.451	16.402	16.354	16.307	16.262	16.217	16.174
500	16.964	16.911	16.860	16.810	16.762	16.715	16.669	16.625
520	17.417	17.362	17.310	17.258	17.208	17.160	17.113	17.067
540	17.851	17.794	17.740	17.687	17.636	17.586	17.537	17.490
560	18.268	18.210	18.154	18.099	18.046	17.995	17.945	17.896
580	18.663	18.603	18.545	18.489	18.434	18.382	18.330	18.281
600	19.041	18.979	18.920	18.862	18.806	18.752	18.700	18.649
620	19.401	19.338	19.277	19.217	19.160	19.105	19.051	18.999
640	19.745	19.680	19.617	19.556	19.498	19.441	19.386	19.332
660	20.078	20.011	19.947	19.884	19.824	19.766	19.710	19.655
680	20.405	20.337	20.271	20.207	20.146	20.087	20.029	19.973
700	20.724	20.654	20.587	20.522	20.459	20.398	20.339	20.282
720	21.043	20.971	20.903	20.836	20.772	20.710	20.650	20.592
740	21.362	21.289	21.219	21.151	21.086	21.022	20.961	20.902
760	21.677	21.602	21.531	21.462	21.395	21.330	21.268	21.208
780	22.000	21.924	21.851	21.780	21.712	21.647	21.583	21.522
800	22.320	22.243	22.169	22.098	22.029	21.962	21.897	21.835
820	22.651	22.573	22.498	22.426	22.356	22.289	22.223	22.160
840	22.975	22.896	22.821	22.748	22.677	22.609	22.543	22.480
860	23.295	23.216	23.140	23.066	22.996	22.928	22.862	22.798
880	23.610	23.531	23.455	23.382	23.312	23.244	23.178	23.115
900	23.917	23.838	23.763	23.691	23.622	23.555	23.491	23.429
920	24.211	24.134	24.061	23.990	23.923	23.858	23.796	23.736
940	24.489	24.414	24.343	24.275	24.210	24.148	24.089	24.032
960	24.742	24.671	24.603	24.539	24.478	24.420	24.365	24.313
980	24.963	24.896	24.834	24.775	24.719	24.666	24.617	24.570
1000	25.142	25.082	25.026	24.973	24.925	24.879	24.836	24.797

TABLE 3
ENTROPY IN CALORIES PER MOLE PER DEGREE KELVIN

TEMP	PRESSURE IN BARS							
	4100	4200	4300	4400	4500	4600	4700	4800
20	0.570	0.552	0.533	0.515	0.497	0.478	0.460	0.442
40	1.741	1.723	1.706	1.688	1.670	1.653	1.635	1.617
60	2.798	2.780	2.763	2.745	2.727	2.710	2.692	2.674
80	3.756	3.738	3.720	3.702	3.684	3.665	3.647	3.629
100	4.645	4.626	4.607	4.588	4.569	4.550	4.532	4.513
120	5.480	5.460	5.440	5.420	5.400	5.381	5.361	5.342
140	6.272	6.251	6.230	6.209	6.188	6.168	6.147	6.127
160	7.037	7.014	6.992	6.971	6.949	6.928	6.906	6.885
180	7.772	7.748	7.725	7.702	7.680	7.658	7.635	7.614
200	8.475	8.451	8.427	8.403	8.380	8.357	8.334	8.311
220	9.155	9.130	9.105	9.080	9.056	9.031	9.008	8.984
240	9.817	9.791	9.764	9.739	9.713	9.688	9.664	9.639
260	10.454	10.426	10.399	10.372	10.346	10.320	10.294	10.269
280	11.063	11.034	11.006	10.978	10.951	10.924	10.897	10.871
300	11.637	11.607	11.578	11.549	11.521	11.493	11.465	11.438
320	12.193	12.162	12.131	12.101	12.072	12.043	12.014	11.986
340	12.736	12.704	12.672	12.641	12.611	12.581	12.551	12.522
360	13.263	13.230	13.197	13.164	13.133	13.102	13.071	13.041
380	13.799	13.764	13.730	13.697	13.664	13.631	13.599	13.568
400	14.306	14.269	14.234	14.199	14.165	14.132	14.099	14.067
420	14.771	14.734	14.697	14.661	14.626	14.591	14.557	14.524
440	15.227	15.188	15.150	15.113	15.077	15.041	15.006	14.971
460	15.684	15.644	15.605	15.567	15.529	15.492	15.456	15.420
480	16.132	16.091	16.050	16.011	15.972	15.934	15.897	15.860
500	16.581	16.539	16.497	16.456	16.417	16.377	16.339	16.302
520	17.022	16.978	16.935	16.894	16.853	16.813	16.773	16.735
540	17.444	17.399	17.355	17.312	17.270	17.229	17.189	17.149
560	17.849	17.803	17.758	17.714	17.671	17.629	17.588	17.548
580	18.232	18.185	18.139	18.094	18.050	18.007	17.965	17.924
600	18.599	18.551	18.504	18.458	18.413	18.369	18.326	18.284
620	18.948	18.899	18.850	18.804	18.758	18.713	18.669	18.626
640	19.281	19.230	19.181	19.133	19.086	19.041	18.996	18.952
660	19.602	19.551	19.501	19.452	19.404	19.358	19.312	19.268
680	19.919	19.866	19.815	19.765	19.717	19.669	19.623	19.578
700	20.227	20.173	20.121	20.070	20.020	19.972	19.925	19.879
720	20.535	20.480	20.427	20.375	20.325	20.275	20.227	20.180
740	20.844	20.788	20.734	20.681	20.629	20.579	20.530	20.483
760	21.149	21.092	21.037	20.983	20.930	20.879	20.830	20.781
780	21.462	21.404	21.348	21.293	21.240	21.188	21.138	21.089
800	21.775	21.716	21.659	21.603	21.550	21.497	21.446	21.396
820	22.099	22.040	21.982	21.926	21.872	21.819	21.767	21.717
840	22.418	22.359	22.301	22.245	22.190	22.137	22.086	22.035
860	22.737	22.677	22.620	22.564	22.509	22.457	22.405	22.355
880	23.055	22.996	22.939	22.883	22.830	22.778	22.727	22.678
900	23.369	23.312	23.256	23.202	23.150	23.099	23.050	23.002
920	23.678	23.623	23.569	23.517	23.467	23.419	23.372	23.326
940	23.978	23.925	23.875	23.826	23.779	23.734	23.690	23.647
960	24.262	24.214	24.168	24.124	24.081	24.040	24.000	23.961
980	24.525	24.483	24.442	24.403	24.366	24.330	24.295	24.262
1000	24.759	24.724	24.690	24.659	24.628	24.599	24.572	24.544

TABLE 3
ENTROPY IN CALORIES PER MOLE PER DEGREE KELVIN

TEMP	PRESSURE IN BARS 4900	5000	5100	5200	5300	5400	5500	5600
20	0.424	0.406	0.389	0.371	0.353	0.336	0.319	0.301
40	1.600	1.582	1.565	1.548	1.530	1.513	1.496	1.479
60	2.657	2.639	2.621	2.604	2.586	2.569	2.552	2.535
80	3.611	3.593	3.575	3.557	3.539	3.522	3.504	3.486
100	4.494	4.476	4.457	4.439	4.420	4.402	4.384	4.366
120	5.322	5.303	5.284	5.265	5.246	5.227	5.208	5.189
140	6.107	6.087	6.067	6.047	6.027	6.008	5.988	5.969
160	6.864	6.844	6.823	6.803	6.782	6.762	6.742	6.722
180	7.592	7.570	7.549	7.528	7.507	7.486	7.465	7.445
200	8.288	8.266	8.244	8.222	8.200	8.179	8.158	8.136
220	8.961	8.938	8.915	8.892	8.870	8.848	8.826	8.804
240	9.615	9.591	9.567	9.544	9.521	9.498	9.476	9.453
260	10.244	10.219	10.195	10.171	10.147	10.123	10.100	10.077
280	10.845	10.819	10.794	10.769	10.745	10.720	10.696	10.673
300	11.411	11.385	11.359	11.333	11.308	11.283	11.258	11.233
320	11.958	11.931	11.904	11.877	11.851	11.825	11.799	11.774
340	12.493	12.465	12.437	12.409	12.382	12.356	12.329	12.303
360	13.011	12.982	12.953	12.924	12.896	12.869	12.841	12.815
380	13.537	13.507	13.477	13.448	13.419	13.390	13.362	13.335
400	14.035	14.003	13.973	13.942	13.912	13.883	13.854	13.825
420	14.491	14.458	14.427	14.395	14.365	14.334	14.304	14.275
440	14.937	14.904	14.871	14.839	14.808	14.776	14.746	14.715
460	15.386	15.351	15.318	15.284	15.252	15.220	15.188	15.157
480	15.824	15.789	15.754	15.720	15.687	15.654	15.622	15.590
500	16.265	16.229	16.193	16.158	16.124	16.090	16.057	16.025
520	16.697	16.660	16.624	16.588	16.553	16.519	16.485	16.451
540	17.111	17.073	17.036	16.999	16.963	16.928	16.894	16.860
560	17.508	17.470	17.432	17.394	17.358	17.322	17.287	17.252
580	17.884	17.844	17.805	17.767	17.730	17.693	17.657	17.622
600	18.243	18.203	18.163	18.125	18.087	18.049	18.013	17.977
620	18.585	18.544	18.503	18.464	18.425	18.387	18.350	18.314
640	18.910	18.868	18.827	18.787	18.748	18.709	18.671	18.634
660	19.224	19.182	19.140	19.099	19.059	19.020	18.981	18.944
680	19.533	19.490	19.448	19.406	19.365	19.325	19.286	19.248
700	19.834	19.790	19.747	19.704	19.663	19.622	19.583	19.543
720	20.135	20.090	20.046	20.003	19.961	19.920	19.879	19.839
740	20.436	20.390	20.346	20.302	20.259	20.217	20.176	20.136
760	20.734	20.687	20.642	20.598	20.554	20.512	20.470	20.429
780	21.040	20.993	20.948	20.903	20.859	20.816	20.773	20.732
800	21.347	21.300	21.254	21.208	21.164	21.120	21.078	21.036
820	21.668	21.620	21.574	21.528	21.483	21.439	21.396	21.354
840	21.986	21.938	21.891	21.846	21.801	21.757	21.714	21.672
860	22.306	22.259	22.212	22.167	22.122	22.079	22.036	21.994
880	22.630	22.583	22.537	22.492	22.448	22.405	22.363	22.322
900	22.955	22.910	22.865	22.821	22.779	22.737	22.696	22.656
920	23.281	23.237	23.195	23.153	23.112	23.072	23.033	22.994
940	23.605	23.564	23.524	23.485	23.447	23.409	23.372	23.335
960	23.923	23.886	23.850	23.814	23.779	23.745	23.711	23.677
980	24.229	24.197	24.166	24.135	24.104	24.074	24.043	24.013
1000	24.518	24.492	24.467	24.441	24.416	24.391	24.365	24.340

TABLE 3
ENTROPY IN CALORIES PER MOLE PER DEGREE KELVIN

TEMP	PRESSURE IN BARS 5700	5800	5900	6000	6100	6200	6300	6400
20	0.284	0.267	0.251	0.234	0.217	0.201	0.185	0.169
40	1.462	1.445	1.428	1.412	1.395	1.379	1.362	1.346
60	2.517	2.500	2.483	2.466	2.449	2.433	2.416	2.400
80	3.469	3.451	3.434	3.416	3.399	3.382	3.365	3.348
100	4.347	4.329	4.311	4.294	4.276	4.258	4.241	4.223
120	5.170	5.152	5.133	5.115	5.097	5.079	5.060	5.042
140	5.950	5.931	5.912	5.893	5.874	5.855	5.837	5.818
160	6.702	6.682	6.663	6.643	6.624	6.605	6.586	6.567
180	7.424	7.404	7.384	7.364	7.344	7.324	7.305	7.285
200	8.115	8.095	8.074	8.053	8.033	8.013	7.993	7.973
220	8.782	8.761	8.740	8.719	8.698	8.677	8.657	8.636
240	9.431	9.409	9.387	9.366	9.344	9.323	9.302	9.281
260	10.054	10.032	10.009	9.987	9.965	9.943	9.922	9.900
280	10.649	10.626	10.603	10.580	10.557	10.535	10.513	10.491
300	11.209	11.185	11.161	11.138	11.115	11.092	11.069	11.047
320	11.749	11.725	11.700	11.676	11.652	11.629	11.605	11.582
340	12.277	12.252	12.227	12.202	12.177	12.153	12.129	12.106
360	12.788	12.762	12.736	12.710	12.685	12.660	12.636	12.611
380	13.307	13.280	13.254	13.227	13.201	13.176	13.150	13.125
400	13.797	13.769	13.742	13.715	13.688	13.662	13.636	13.610
420	14.246	14.217	14.189	14.161	14.134	14.107	14.080	14.054
440	14.686	14.656	14.627	14.599	14.571	14.543	14.515	14.488
460	15.127	15.096	15.067	15.037	15.009	14.980	14.952	14.924
480	15.558	15.528	15.497	15.467	15.438	15.408	15.380	15.351
500	15.992	15.961	15.930	15.899	15.869	15.839	15.810	15.781
520	16.418	16.386	16.354	16.323	16.292	16.262	16.232	16.202
540	16.826	16.793	16.761	16.729	16.697	16.666	16.636	16.606
560	17.218	17.184	17.151	17.119	17.087	17.055	17.024	16.993
580	17.587	17.553	17.520	17.486	17.454	17.422	17.390	17.359
600	17.942	17.907	17.873	17.839	17.806	17.773	17.741	17.710
620	18.278	18.242	18.208	18.173	18.140	18.107	18.074	18.042
640	18.597	18.561	18.526	18.491	18.457	18.424	18.391	18.358
660	18.906	18.870	18.834	18.799	18.764	18.730	18.697	18.664
680	19.210	19.173	19.137	19.101	19.066	19.031	18.997	18.964
700	19.505	19.467	19.431	19.394	19.359	19.323	19.289	19.255
720	19.800	19.762	19.725	19.688	19.652	19.616	19.581	19.547
740	20.096	20.058	20.020	19.982	19.945	19.909	19.874	19.839
760	20.389	20.350	20.311	20.273	20.236	20.199	20.163	20.128
780	20.691	20.652	20.613	20.574	20.536	20.499	20.463	20.427
800	20.995	20.955	20.915	20.876	20.838	20.801	20.764	20.728
820	21.313	21.273	21.233	21.194	21.156	21.118	21.081	21.044
840	21.631	21.590	21.550	21.511	21.473	21.435	21.398	21.361
860	21.953	21.912	21.873	21.834	21.795	21.757	21.720	21.684
880	22.281	22.241	22.202	22.163	22.125	22.088	22.051	22.014
900	22.616	22.577	22.538	22.500	22.463	22.426	22.389	22.353
920	22.956	22.918	22.880	22.844	22.807	22.771	22.735	22.700
940	23.299	23.263	23.228	23.192	23.157	23.122	23.087	23.053
960	23.643	23.610	23.576	23.543	23.510	23.476	23.443	23.409
980	23.983	23.953	23.922	23.891	23.860	23.829	23.797	23.765
1000	24.314	24.287	24.260	24.232	24.204	24.175	24.145	24.115

TABLE 3
ENTROPY IN CALORIES PER MOLE PER DEGREE KELVIN

	PRESSURE IN BARS							
TEMP	6500	6600	6700	6800	6900	7000	7100	7200
20	0.153	0.137	0.121	0.105	0.089	0.074	0.058	0.043
40	1.330	1.314	1.298	1.282	1.266	1.250	1.235	1.219
60	2.383	2.367	2.350	2.334	2.318	2.302	2.286	2.270
80	3.331	3.314	3.297	3.281	3.264	3.248	3.231	3.215
100	4.206	4.188	4.171	4.154	4.137	4.120	4.103	4.087
120	5.025	5.007	4.989	4.972	4.954	4.937	4.919	4.902
140	5.800	5.781	5.763	5.745	5.727	5.710	5.692	5.674
160	6.548	6.529	6.511	6.492	6.474	6.456	6.437	6.419
180	7.266	7.247	7.228	7.209	7.190	7.171	7.153	7.134
200	7.953	7.934	7.914	7.895	7.875	7.856	7.837	7.819
220	8.616	8.596	8.576	8.556	8.537	8.517	8.498	8.479
240	9.260	9.240	9.219	9.199	9.179	9.159	9.140	9.120
260	9.879	9.858	9.837	9.817	9.796	9.776	9.756	9.736
280	10.469	10.448	10.426	10.405	10.384	10.364	10.343	10.323
300	11.024	11.002	10.980	10.959	10.937	10.916	10.895	10.874
320	11.559	11.537	11.514	11.492	11.470	11.449	11.427	11.406
340	12.082	12.059	12.036	12.013	11.991	11.968	11.946	11.925
360	12.587	12.563	12.540	12.517	12.493	12.471	12.448	12.426
380	13.101	13.076	13.052	13.028	13.005	12.981	12.958	12.935
400	13.585	13.560	13.535	13.511	13.486	13.463	13.439	13.416
420	14.028	14.002	13.977	13.952	13.927	13.903	13.878	13.854
440	14.462	14.435	14.409	14.384	14.358	14.333	14.309	14.284
460	14.897	14.870	14.844	14.817	14.791	14.766	14.741	14.716
480	15.324	15.296	15.269	15.242	15.216	15.190	15.164	15.138
500	15.752	15.724	15.697	15.669	15.642	15.616	15.589	15.564
520	16.173	16.145	16.117	16.089	16.061	16.034	16.007	15.981
540	16.576	16.547	16.518	16.490	16.462	16.434	16.407	16.381
560	16.963	16.934	16.904	16.876	16.847	16.819	16.792	16.764
580	17.329	17.299	17.269	17.240	17.211	17.182	17.154	17.127
600	17.679	17.648	17.618	17.588	17.559	17.530	17.502	17.474
620	18.011	17.979	17.949	17.919	17.889	17.860	17.831	17.803
640	18.326	18.295	18.264	18.233	18.203	18.173	18.144	18.115
660	18.631	18.599	18.568	18.537	18.506	18.476	18.447	18.418
680	18.931	18.898	18.867	18.835	18.804	18.774	18.744	18.714
700	19.222	19.189	19.156	19.125	19.093	19.063	19.032	19.002
720	19.513	19.479	19.447	19.415	19.383	19.352	19.321	19.291
740	19.805	19.771	19.738	19.705	19.673	19.641	19.610	19.580
760	20.093	20.059	20.026	19.993	19.960	19.928	19.897	19.866
780	20.392	20.358	20.324	20.290	20.257	20.225	20.193	20.162
800	20.692	20.657	20.623	20.589	20.556	20.523	20.491	20.459
820	21.009	20.973	20.939	20.905	20.871	20.838	20.805	20.773
840	21.325	21.289	21.255	21.220	21.186	21.153	21.120	21.088
860	21.647	21.612	21.577	21.542	21.508	21.474	21.441	21.409
880	21.978	21.943	21.907	21.873	21.839	21.805	21.771	21.738
900	22.317	22.282	22.247	22.212	22.178	22.144	22.110	22.077
920	22.664	22.629	22.595	22.560	22.526	22.491	22.457	22.424
940	23.018	22.984	22.949	22.915	22.880	22.846	22.812	22.777
960	23.376	23.342	23.308	23.274	23.239	23.205	23.170	23.135
980	23.732	23.699	23.666	23.632	23.597	23.562	23.527	23.491
1000	24.083	24.051	24.018	23.984	23.949	23.913	23.877	23.839

TABLE 3
ENTROPY IN CALORIES PER MOLE PER DEGREE KELVIN

TEMP	PRESSURE IN BARS							
	7300	7400	7500	7600	7700	7800	7900	8000
20	0.027	0.011	-0.004	-0.020	-0.035	-0.051	-0.067	-0.083
40	1.203	1.188	1.172	1.157	1.141	1.125	1.110	1.094
60	2.254	2.238	2.223	2.207	2.191	2.175	2.160	2.144
80	3.199	3.183	3.166	3.150	3.134	3.118	3.103	3.087
100	4.070	4.053	4.037	4.020	4.004	3.988	3.972	3.955
120	4.885	4.868	4.851	4.834	4.818	4.801	4.785	4.768
140	5.657	5.639	5.622	5.605	5.588	5.571	5.554	5.537
160	6.401	6.384	6.366	6.348	6.331	6.314	6.296	6.279
180	7.116	7.098	7.080	7.062	7.044	7.026	7.009	6.991
200	7.800	7.781	7.763	7.745	7.726	7.708	7.690	7.673
220	8.460	8.441	8.422	8.403	8.385	8.366	8.348	8.330
240	9.101	9.081	9.062	9.043	9.024	9.005	8.987	8.968
260	9.716	9.696	9.677	9.657	9.638	9.619	9.600	9.581
280	10.302	10.282	10.262	10.243	10.223	10.204	10.184	10.165
300	10.854	10.833	10.813	10.793	10.773	10.753	10.733	10.714
320	11.385	11.364	11.343	11.322	11.302	11.282	11.262	11.242
340	11.903	11.882	11.860	11.839	11.819	11.798	11.778	11.757
360	12.404	12.382	12.360	12.339	12.318	12.297	12.276	12.255
380	12.913	12.890	12.868	12.847	12.825	12.803	12.782	12.761
400	13.392	13.370	13.347	13.325	13.303	13.281	13.259	13.238
420	13.831	13.808	13.785	13.762	13.739	13.717	13.695	13.673
440	14.260	14.236	14.213	14.190	14.167	14.144	14.121	14.099
460	14.691	14.667	14.643	14.619	14.596	14.573	14.550	14.527
480	15.113	15.089	15.064	15.040	15.016	14.993	14.969	14.946
500	15.538	15.513	15.488	15.463	15.439	15.415	15.392	15.368
520	15.955	15.929	15.904	15.879	15.855	15.830	15.806	15.783
540	16.354	16.328	16.302	16.277	16.252	16.227	16.203	16.179
560	16.738	16.711	16.685	16.659	16.634	16.609	16.584	16.560
580	17.099	17.073	17.046	17.020	16.994	16.969	16.944	16.919
600	17.446	17.419	17.392	17.365	17.339	17.314	17.288	17.263
620	17.775	17.747	17.720	17.693	17.667	17.641	17.615	17.590
640	18.087	18.059	18.032	18.005	17.978	17.951	17.925	17.900
660	18.389	18.361	18.333	18.305	18.278	18.252	18.225	18.199
680	18.685	18.657	18.628	18.601	18.573	18.546	18.520	18.493
700	18.973	18.944	18.915	18.887	18.859	18.832	18.805	18.779
720	19.261	19.231	19.203	19.174	19.146	19.118	19.091	19.064
740	19.549	19.520	19.491	19.462	19.433	19.405	19.378	19.351
760	19.835	19.805	19.775	19.746	19.718	19.689	19.661	19.634
780	20.131	20.100	20.070	20.041	20.012	19.983	19.955	19.927
800	20.428	20.397	20.367	20.337	20.308	20.279	20.251	20.222
820	20.742	20.710	20.680	20.650	20.620	20.591	20.562	20.533
840	21.056	21.024	20.993	20.963	20.932	20.903	20.873	20.845
860	21.376	21.344	21.313	21.282	21.251	21.221	21.191	21.162
880	21.706	21.673	21.641	21.610	21.579	21.548	21.517	21.487
900	22.044	22.011	21.979	21.946	21.914	21.883	21.852	21.821
920	22.390	22.357	22.323	22.290	22.258	22.225	22.193	22.160
940	22.743	22.709	22.675	22.641	22.607	22.573	22.539	22.505
960	23.100	23.064	23.029	22.993	22.957	22.921	22.885	22.849
980	23.454	23.417	23.380	23.342	23.304	23.265	23.227	23.187
1000	23.801	23.762	23.722	23.681	23.640	23.598	23.555	23.511

TABLE 3
ENTROPY IN CALORIES PER MOLE PER DEGREE KELVIN

TEMP	PRESSURE IN BARS 8100	8200	8300	8400	8500	8600	8700	8800
20	-0.099	-0.115	-0.132	-0.148	-0.165	-0.182	-0.199	-0.216
40	1.078	1.063	1.047	1.031	1.015	0.999	0.982	0.966
60	2.128	2.112	2.097	2.081	2.065	2.049	2.034	2.018
80	3.071	3.055	3.039	3.023	3.008	2.992	2.976	2.961
100	3.939	3.923	3.907	3.891	3.876	3.860	3.844	3.828
120	4.752	4.735	4.719	4.703	4.687	4.671	4.655	4.639
140	5.520	5.504	5.487	5.471	5.454	5.438	5.422	5.406
160	6.262	6.245	6.228	6.211	6.195	6.178	6.162	6.145
180	6.974	6.956	6.939	6.922	6.905	6.888	6.872	6.855
200	7.655	7.637	7.620	7.602	7.585	7.568	7.551	7.534
220	8.312	8.294	8.276	8.258	8.241	8.223	8.206	8.188
240	8.950	8.932	8.913	8.895	8.878	8.860	8.842	8.825
260	9.562	9.544	9.525	9.507	9.489	9.471	9.453	9.435
280	10.146	10.127	10.108	10.090	10.071	10.053	10.035	10.017
300	10.694	10.675	10.656	10.637	10.618	10.600	10.581	10.563
320	11.222	11.203	11.183	11.164	11.145	11.126	11.107	11.089
340	11.737	11.718	11.698	11.678	11.659	11.640	11.621	11.602
360	12.235	12.215	12.195	12.175	12.155	12.136	12.116	12.097
380	12.740	12.720	12.700	12.679	12.659	12.640	12.620	12.601
400	13.217	13.196	13.175	13.154	13.134	13.114	13.094	13.075
420	13.652	13.630	13.609	13.588	13.568	13.547	13.527	13.507
440	14.077	14.056	14.034	14.013	13.992	13.972	13.951	13.931
460	14.505	14.483	14.461	14.440	14.418	14.397	14.377	14.356
480	14.924	14.901	14.879	14.857	14.836	14.815	14.794	14.773
500	15.345	15.322	15.300	15.278	15.256	15.234	15.213	15.192
520	15.759	15.736	15.713	15.691	15.669	15.647	15.625	15.604
540	16.155	16.132	16.109	16.086	16.063	16.041	16.019	15.997
560	16.536	16.512	16.488	16.465	16.442	16.420	16.398	16.376
580	16.895	16.871	16.847	16.823	16.800	16.777	16.755	16.733
600	17.238	17.214	17.190	17.166	17.143	17.120	17.097	17.074
620	17.564	17.540	17.515	17.491	17.468	17.444	17.421	17.398
640	17.874	17.849	17.825	17.800	17.776	17.753	17.729	17.706
660	18.174	18.148	18.123	18.099	18.074	18.050	18.027	18.003
680	18.467	18.442	18.417	18.392	18.367	18.343	18.319	18.295
700	18.752	18.727	18.701	18.676	18.651	18.627	18.603	18.579
720	19.038	19.012	18.986	18.961	18.936	18.911	18.886	18.862
740	19.324	19.298	19.272	19.246	19.221	19.196	19.171	19.147
760	19.607	19.580	19.554	19.528	19.503	19.478	19.453	19.428
780	19.900	19.873	19.847	19.820	19.795	19.769	19.744	19.719
800	20.195	20.168	20.141	20.114	20.088	20.063	20.037	20.012
820	20.505	20.478	20.451	20.424	20.397	20.371	20.346	20.321
840	20.816	20.788	20.761	20.733	20.707	20.680	20.654	20.628
860	21.133	21.105	21.076	21.049	21.021	20.994	20.967	20.941
880	21.458	21.428	21.399	21.371	21.343	21.315	21.287	21.260
900	21.790	21.760	21.730	21.700	21.671	21.642	21.613	21.584
920	22.128	22.097	22.065	22.034	22.003	21.972	21.942	21.912
940	22.471	22.437	22.404	22.370	22.337	22.304	22.271	22.239
960	22.813	22.777	22.740	22.704	22.668	22.632	22.596	22.560
980	23.148	23.108	23.068	23.028	22.987	22.947	22.906	22.866
1000	23.467	23.423	23.378	23.332	23.286	23.240	23.193	23.146

TABLE 3
ENTROPY IN CALORIES PER MOLE PER DEGREE KELVIN

TEMP	PRESSURE IN BARS							
	8900	9000	9100	9200	9300	9400	9500	9600
20	(-0.233)	(-0.251)	(-0.269)	(-0.287)	(-0.305)	(-0.324)	(-0.342)	(-0.361)
40	0.950	0.933	0.916	0.900	0.883	0.866	0.849	0.832
60	2.002	1.986	1.970	1.954	1.938	1.922	1.906	1.890
80	2.945	2.929	2.914	2.898	2.883	2.867	2.852	2.836
100	3.813	3.797	3.782	3.766	3.751	3.735	3.720	3.705
120	4.623	4.608	4.592	4.576	4.561	4.546	4.530	4.515
140	5.390	5.374	5.358	5.342	5.327	5.311	5.296	5.281
160	6.129	6.113	6.097	6.081	6.065	6.049	6.034	6.018
180	6.838	6.822	6.806	6.789	6.773	6.757	6.741	6.726
200	7.517	7.500	7.484	7.467	7.451	7.434	7.418	7.402
220	8.171	8.154	8.137	8.120	8.104	8.087	8.071	8.054
240	8.807	8.790	8.773	8.756	8.739	8.722	8.705	8.688
260	9.417	9.400	9.382	9.365	9.348	9.331	9.314	9.297
280	9.999	9.981	9.963	9.945	9.928	9.911	9.893	9.876
300	10.545	10.527	10.509	10.491	10.473	10.456	10.438	10.421
320	11.070	11.052	11.034	11.016	10.998	10.980	10.963	10.945
340	11.583	11.565	11.546	11.528	11.510	11.492	11.474	11.457
360	12.078	12.059	12.041	12.022	12.004	11.986	11.968	11.950
380	12.581	12.562	12.544	12.525	12.507	12.488	12.470	12.452
400	13.055	13.036	13.017	12.998	12.979	12.961	12.943	12.925
420	13.488	13.468	13.449	13.430	13.411	13.392	13.374	13.356
440	13.911	13.891	13.872	13.852	13.833	13.815	13.796	13.778
460	14.336	14.316	14.296	14.277	14.257	14.238	14.220	14.201
480	14.752	14.732	14.712	14.692	14.673	14.654	14.635	14.616
500	15.171	15.151	15.130	15.110	15.091	15.071	15.052	15.033
520	15.583	15.562	15.541	15.521	15.501	15.481	15.462	15.442
540	15.976	15.955	15.934	15.913	15.893	15.873	15.853	15.834
560	16.354	16.333	16.311	16.291	16.270	16.250	16.230	16.210
580	16.711	16.689	16.667	16.646	16.625	16.605	16.584	16.564
600	17.052	17.030	17.008	16.987	16.966	16.945	16.924	16.904
620	17.376	17.353	17.331	17.310	17.288	17.267	17.246	17.225
640	17.683	17.661	17.638	17.616	17.595	17.573	17.552	17.531
660	17.980	17.957	17.935	17.913	17.891	17.869	17.847	17.826
680	18.272	18.249	18.226	18.204	18.181	18.159	18.138	18.116
700	18.555	18.532	18.509	18.486	18.463	18.441	18.419	18.398
720	18.839	18.815	18.792	18.769	18.746	18.724	18.702	18.680
740	19.123	19.099	19.076	19.053	19.030	19.007	18.985	18.963
760	19.404	19.380	19.357	19.333	19.310	19.287	19.265	19.243
780	19.695	19.671	19.647	19.624	19.601	19.578	19.555	19.533
800	19.988	19.963	19.939	19.916	19.892	19.869	19.847	19.824
820	20.296	20.271	20.247	20.223	20.199	20.176	20.153	20.130
840	20.603	20.578	20.553	20.529	20.505	20.482	20.458	20.435
860	20.915	20.890	20.864	20.840	20.815	20.791	20.767	20.744
880	21.233	21.207	21.181	21.155	21.130	21.105	21.080	21.056
900	21.556	21.529	21.501	21.474	21.448	21.422	21.396	21.370
920	21.882	21.852	21.823	21.795	21.766	21.738	21.710	21.683
940	22.207	22.175	22.143	22.111	22.080	22.049	22.019	21.989
960	22.524	22.488	22.453	22.418	22.383	22.348	22.314	22.280
980	22.825	22.784	22.744	22.703	22.663	22.623	22.583	22.544
1000	23.099	23.052	23.004	22.957	22.909	22.862	22.815	22.768

TABLE 3
ENTROPY IN CALORIES PER MOLE PER DEGREE KELVIN

TEMP	PRESSURE IN BARS 9700	9800	9900	10000
20	(-0.380)	(-0.399)	(-0.419)	(-0.438)
40	0.815	0.797	0.780	0.763
60	1.874	1.858	1.842	1.826
80	2.821	2.806	2.791	2.775
100	3.690	3.675	3.660	3.645
120	4.500	4.485	4.471	4.456
140	5.265	5.250	5.235	5.221
160	6.003	5.988	5.972	5.957
180	6.710	6.694	6.679	6.664
200	7.386	7.370	7.354	7.339
220	8.038	8.022	8.006	7.990
240	8.672	8.655	8.639	8.623
260	9.280	9.263	9.247	9.230
280	9.859	9.843	9.826	9.809
300	10.404	10.387	10.370	10.353
320	10.928	10.911	10.893	10.876
340	11.439	11.422	11.405	11.388
360	11.933	11.915	11.898	11.881
380	12.435	12.417	12.400	12.382
400	12.907	12.889	12.872	12.854
420	13.338	13.320	13.303	13.285
440	13.760	13.742	13.724	13.707
460	14.183	14.165	14.147	14.129
480	14.597	14.579	14.561	14.543
500	15.014	14.996	14.978	14.960
520	15.424	15.405	15.386	15.368
540	15.815	15.796	15.777	15.759
560	16.190	16.171	16.152	16.134
580	16.545	16.525	16.506	16.487
600	16.884	16.864	16.844	16.825
620	17.205	17.185	17.165	17.145
640	17.510	17.490	17.470	17.450
660	17.805	17.784	17.764	17.744
680	18.095	18.074	18.053	18.032
700	18.376	18.355	18.334	18.313
720	18.658	18.637	18.615	18.594
740	18.941	18.919	18.898	18.877
760	19.221	19.199	19.178	19.156
780	19.511	19.489	19.467	19.446
800	19.802	19.780	19.758	19.737
820	20.108	20.086	20.064	20.043
840	20.413	20.390	20.369	20.347
860	20.721	20.698	20.676	20.654
880	21.032	21.009	20.986	20.963
900	21.345	21.321	21.297	21.273
920	21.656	21.630	21.604	21.578
940	21.959	21.930	21.901	21.873
960	22.246	22.213	22.180	22.148
980	22.505	22.466	22.427	22.390
1000	22.721	22.675	22.629	22.583

TABLE 4
ENTHALPY IN CALORIES PER MOLE

TEMP	PRESSURE IN BARS							
	100	200	300	400	500	600	700	800
20	402	442	481	520	559	598	637	675
40	759	797	834	871	909	946	982	1019
60	1117	1153	1189	1225	1260	1296	1331	1367
80	1476	1510	1544	1579	1613	1647	1682	1716
100	1836	1868	1901	1934	1967	1999	2032	2065
120	2198	2228	2259	2289	2320	2352	2383	2414
140	2562	2591	2619	2648	2677	2707	2736	2766
160	2931	2957	2983	3010	3038	3065	3093	3122
180	3304	3327	3351	3376	3401	3427	3453	3479
200	3684	3703	3723	3745	3767	3791	3814	3839
220	4071	4086	4102	4120	4139	4159	4180	4202
240	4469	4477	4488	4501	4516	4533	4551	4570
260	4880	4879	4883	4890	4900	4913	4927	4942
280	5315	5299	5292	5291	5294	5300	5309	5320
300	5779	5745	5719	5706	5698	5693	5698	5702
320	11980	6223	6171	6136	6119	6102	6093	6093
340	12410	6769	6661	6597	6558	6528	6507	6489
360	12754	7488	7216	7096	7023	6971	6937	6911
380	13051	11451	7918	7655	7531	7449	7393	7354
400	13327	12130	9282	8327	8086	7952	7871	7810
420	13585	12608	10994	9213	8710	8490	8357	8266
440	13826	13000	11838	10310	9445	9080	8877	8744
460	14062	13340	12414	11253	10254	9725	9437	9252
480	14290	13649	12875	11950	11037	10401	10022	9781
500	14514	13938	13271	12505	11713	11063	10620	10327
520	14738	14217	13628	12974	12290	11674	11205	10870
540	14953	14480	13955	13383	12784	12225	11756	11399
560	15168	14734	14256	13748	13223	12716	12268	11907
580	15383	14979	14544	14088	13619	13159	12741	12384
600	15594	15220	14824	14407	13981	13563	13176	12836
620	15805	15457	15091	14708	14325	13942	13585	13258
640	16016	15689	15349	15000	14648	14299	13963	13658
660	16227	15921	15603	15280	14953	14635	14325	14041
680	16433	16149	15852	15556	15254	14957	14669	14402
700	16644	16377	16102	15822	15543	15267	15000	14751
720	16855	16601	16343	16085	15822	15568	15323	15086
740	17066	16825	16584	16343	16102	15861	15633	15409
760	17277	17053	16825	16597	16373	16149	15934	15728
780	17487	17277	17062	16851	16640	16429	16231	16037
800	17698	17500	17298	17100	16902	16709	16519	16339

TABLE 4
ENTHALPY IN CALORIES PER MOLE

TEMP	PRESSURE IN BARS 900	1000	1100	1200	1300	1400	1500	1600
20	712	750	785	820	855	890	925	960
40	1056	1092	1128	1164	1200	1236	1271	1307
60	1402	1438	1473	1509	1544	1580	1615	1650
80	1750	1784	1819	1853	1888	1922	1957	1991
100	2098	2131	2164	2197	2231	2264	2297	2330
120	2446	2477	2509	2540	2572	2604	2636	2668
140	2796	2827	2857	2887	2917	2948	2978	3009
160	3150	3179	3208	3237	3266	3295	3324	3353
180	3507	3534	3561	3588	3616	3643	3671	3699
200	3864	3889	3914	3940	3966	3993	4020	4046
220	4224	4247	4271	4296	4321	4346	4371	4397
240	4590	4611	4633	4655	4679	4702	4726	4751
260	4959	4977	4997	5018	5039	5061	5083	5106
280	5333	5347	5363	5381	5400	5420	5440	5462
300	5706	5715	5727	5742	5758	5775	5793	5812
320	6093	6093	6101	6111	6123	6137	6152	6169
340	6485	6481	6482	6487	6494	6504	6516	6529
360	6894	6881	6874	6873	6875	6880	6887	6897
380	7328	7307	7291	7282	7278	7277	7279	7284
400	7767	7737	7711	7693	7681	7674	7671	7671
420	8202	8159	8126	8102	8084	8070	8061	8055
440	8654	8589	8541	8505	8477	8456	8439	8427
460	9127	9041	8973	8922	8883	8852	8827	8809
480	9618	9501	9412	9344	9291	9250	9217	9191
500	10125	9979	9867	9781	9715	9663	9621	9587
520	10633	10461	10328	10225	10144	10081	10029	9988
540	11136	10934	10781	10662	10568	10493	10432	10383
560	11623	11407	11238	11104	10997	10912	10842	10784
580	12096	11864	11681	11534	11416	11320	11241	11176
600	12548	12307	12114	11957	11830	11725	11638	11565
620	12978	12737	12538	12374	12239	12126	12032	11952
640	13387	13150	12949	12780	12638	12519	12419	12333
660	13778	13550	13349	13178	13033	12909	12804	12713
680	14153	13933	13736	13564	13417	13290	13181	13087
700	14514	14303	14112	13942	13795	13666	13555	13458
720	14867	14660	14476	14310	14164	14035	13923	13824
740	15203	15009	14832	14671	14528	14400	14288	14189
760	15530	15349	15180	15025	14884	14759	14647	14548
780	15852	15676	15515	15365	15229	15105	14995	14897
800	16162	15999	15844	15700	15567	15447	15338	15240
820		16321	16172	16033	15903	15785	15678	15581
840		16629	16485	16350	16223	16107	16001	15905
860		16927	16787	16655	16531	16416	16311	16216
880		17219	17083	16955	16833	16719	16614	16520
900		17505	17397	17296	17201	17112	17029	16951
920		17782	17690	17602	17516	17434	17356	17282
940		18054	17977	17898	17819	17742	17667	17595
960		18319	18250	18176	18101	18026	17952	17880
980		18578	18510	18436	18359	18281	18204	18129
1000		18832	18754	18671	18585	18499	18414	18332

TABLE 4
ENTHALPY IN CALORIES PER MOLE

TEMP	PRESSURE IN BARS 1700	1800	1900	2000	2100	2200	2300	2400
20	995	1030	1065	1099	1134	1168	1202	1237
40	1342	1377	1412	1447	1482	1517	1552	1586
60	1685	1720	1755	1790	1825	1859	1894	1928
80	2025	2059	2094	2128	2162	2196	2229	2263
100	2363	2396	2429	2463	2496	2529	2562	2595
120	2700	2732	2763	2795	2827	2859	2892	2924
140	3039	3070	3101	3131	3162	3193	3224	3255
160	3382	3412	3441	3471	3501	3531	3561	3591
180	3727	3756	3784	3813	3841	3870	3899	3928
200	4073	4100	4128	4155	4183	4211	4239	4267
220	4423	4449	4475	4501	4528	4555	4582	4609
240	4775	4800	4825	4850	4876	4902	4928	4954
260	5129	5153	5177	5201	5225	5250	5275	5300
280	5483	5505	5528	5550	5573	5597	5621	5645
300	5832	5852	5873	5894	5915	5938	5960	5983
320	6186	6204	6223	6242	6262	6283	6304	6325
340	6543	6559	6575	6593	6611	6630	6649	6669
360	6908	6920	6934	6949	6965	6982	6999	7017
380	7292	7301	7311	7324	7337	7351	7366	7383
400	7673	7679	7686	7695	7705	7717	7730	7744
420	8053	8054	8057	8062	8057	8065	8076	8087
440	8419	8415	8414	8415	8409	8415	8422	8431
460	8795	8785	8779	8776	8777	8779	8783	8789
480	9171	9156	9145	9137	9148	9146	9147	9150
500	9560	9539	9523	9510	9532	9526	9523	9523
520	9954	9927	9905	9888	9917	9908	9901	9898
540	10342	10309	10281	10259	10293	10280	10270	10263
560	10736	10696	10663	10636	10670	10652	10639	10628
580	11121	11076	11037	11005	11032	11011	10994	10980
600	11504	11452	11409	11372	11387	11362	11341	11324
620	11885	11828	11779	11738	11735	11706	11682	11662
640	12261	12199	12146	12100	12076	12043	12016	11993
660	12636	12569	12512	12463	12413	12378	12347	12322
680	13006	12936	12875	12823	12746	12708	12674	12646
700	13374	13301	13237	13182	13080	13038	13003	12972
720	13738	13663	13597	13539	13414	13371	13333	13300
740	14101	14024	13957	13897	13756	13710	13670	13635
760	14460	14383	14314	14253	14103	14056	14014	13978
780	14809	14731	14662	14600	14453	14405	14362	14324
800	15153	15075	15006	14944	14813	14763	14719	14680
820	15495	15417	15347	15285	15185	15135	15090	15049
840	15819	15741	15671	15610	15553	15503	15457	15416
860	16130	16052	15982	15920	15918	15868	15822	15780
880	16434	16356	16287	16224	16280	16230	16184	16143
900	16878	16810	16747	16688	16634	16585	16539	16497
920	17211	17145	17083	17026	16972	16923	16877	16836
940	17526	17461	17400	17343	17289	17240	17196	17155
960	17811	17745	17684	17626	17573	17524	17480	17440
980	18058	17990	17926	17867	17813	17764	17720	17681
1000	18254	18181	18114	18052	17996	17946	17902	17864

TABLE 4
ENTHALPY IN CALORIES PER MOLE

TEMP	PRESSURE IN BARS 2500	2600	2700	2800	2900	3000	3100	3200
20	1271	1305	1339	1372	1406	1440	1473	1507
40	1621	1655	1690	1724	1758	1792	1826	1860
60	1962	1997	2031	2065	2099	2133	2167	2200
80	2297	2331	2365	2398	2432	2465	2499	2532
100	2628	2661	2694	2727	2760	2793	2826	2859
120	2956	2988	3020	3052	3085	3117	3149	3181
140	3287	3318	3349	3380	3412	3443	3475	3506
160	3621	3651	3682	3712	3743	3773	3804	3835
180	3957	3987	4016	4046	4076	4106	4136	4166
200	4295	4324	4352	4381	4410	4439	4468	4497
220	4636	4664	4692	4719	4748	4776	4804	4833
240	4980	5007	5034	5061	5088	5116	5143	5171
260	5325	5351	5377	5403	5430	5456	5483	5510
280	5669	5694	5719	5744	5770	5795	5821	5847
300	6006	6030	6054	6078	6102	6127	6152	6177
320	6347	6370	6392	6415	6439	6463	6487	6511
340	6689	6710	6732	6754	6776	6798	6821	6845
360	7036	7055	7075	7096	7117	7138	7160	7182
380	7400	7417	7436	7454	7474	7494	7514	7535
400	7758	7774	7791	7808	7826	7844	7863	7883
420	8100	8113	8128	8143	8160	8177	8194	8212
440	8441	8453	8465	8479	8493	8508	8524	8541
460	8797	8806	8816	8827	8840	8853	8868	8883
480	9155	9161	9169	9178	9189	9201	9213	9227
500	9525	9529	9534	9541	9550	9559	9570	9582
520	9897	9898	9901	9906	9912	9920	9929	9939
540	10259	10257	10257	10260	10264	10270	10277	10285
560	10621	10617	10615	10615	10617	10620	10626	10633
580	10970	10963	10958	10956	10956	10957	10961	10966
600	11311	11301	11294	11289	11287	11287	11288	11292
620	11646	11634	11624	11617	11613	11611	11611	11612
640	11974	11959	11947	11938	11932	11927	11925	11925
660	12300	12283	12268	12257	12249	12243	12239	12237
680	12622	12602	12586	12573	12562	12554	12549	12545
700	12946	12924	12905	12890	12877	12867	12860	12855
720	13272	13247	13227	13209	13195	13184	13175	13168
740	13605	13579	13556	13537	13521	13508	13497	13488
760	13946	13918	13894	13873	13855	13840	13828	13817
780	14290	14260	14235	14212	14193	14176	14162	14150
800	14645	14614	14587	14563	14542	14524	14508	14495
820	15013	14981	14952	14927	14905	14886	14869	14855
840	15379	15346	15316	15290	15267	15246	15229	15213
860	15743	15709	15678	15651	15627	15606	15588	15572
880	16105	16070	16040	16012	15988	15967	15948	15933
900	16459	16425	16395	16367	16343	16322	16305	16289
920	16799	16765	16735	16709	16686	16666	16649	16635
940	17118	17086	17057	17032	17011	16993	16978	16967
960	17405	17374	17347	17324	17306	17291	17280	17272
980	17648	17619	17595	17576	17561	17550	17544	17541
1000	17833	17807	17786	17771	17761	17757	17756	17761

TABLE 4
ENTHALPY IN CALORIES PER MOLE

TEMP	PRESSURE IN BARS 3300	3400	3500	3600	3700	3800	3900	4000
20	1540	1574	1607	1640	1673	1706	1739	1772
40	1893	1927	1960	1994	2027	2061	2094	2127
60	2234	2268	2301	2335	2368	2401	2434	2468
80	2566	2599	2632	2665	2698	2731	2764	2797
100	2891	2924	2957	2990	3023	3055	3088	3120
120	3214	3246	3278	3310	3343	3375	3407	3439
140	3538	3570	3601	3633	3665	3696	3728	3760
160	3866	3897	3928	3959	3990	4021	4052	4083
180	4196	4226	4256	4287	4317	4348	4378	4409
200	4527	4556	4586	4616	4646	4676	4706	4736
220	4862	4890	4919	4949	4978	5007	5037	5066
240	5199	5227	5256	5284	5313	5341	5370	5399
260	5537	5565	5592	5620	5648	5676	5704	5732
280	5874	5900	5927	5954	5981	6009	6036	6064
300	6203	6229	6255	6281	6307	6334	6361	6388
320	6536	6560	6586	6611	6637	6662	6688	6715
340	6868	6892	6916	6941	6966	6991	7016	7041
360	7204	7227	7251	7274	7298	7322	7346	7371
380	7557	7578	7601	7623	7646	7669	7693	7716
400	7903	7924	7945	7966	7988	8010	8033	8055
420	8231	8250	8270	8290	8311	8332	8354	8376
440	8558	8576	8595	8614	8633	8653	8674	8695
460	8899	8915	8933	8951	8969	8988	9007	9027
480	9241	9256	9272	9289	9306	9324	9342	9361
500	9595	9609	9623	9638	9654	9671	9688	9706
520	9950	9962	9975	9989	10004	10020	10036	10052
540	10295	10306	10317	10330	10344	10358	10373	10388
560	10641	10650	10660	10671	10684	10697	10711	10725
580	10972	10980	10989	10999	11010	11022	11035	11048
600	11297	11303	11310	11319	11329	11339	11351	11364
620	11615	11620	11626	11634	11642	11652	11662	11674
640	11927	11930	11935	11941	11948	11956	11966	11976
660	12237	12239	12242	12247	12253	12260	12268	12277
680	12543	12544	12545	12549	12553	12559	12566	12574
700	12852	12850	12851	12853	12856	12861	12866	12873
720	13163	13160	13159	13159	13161	13165	13169	13175
740	13482	13478	13475	13474	13475	13477	13480	13485
760	13810	13804	13800	13797	13797	13798	13800	13803
780	14141	14134	14128	14125	14123	14122	14124	14126
800	14485	14476	14469	14465	14462	14460	14460	14462
820	14843	14833	14825	14820	14816	14813	14813	14813
840	15201	15190	15182	15175	15171	15168	15167	15167
860	15559	15548	15539	15533	15528	15525	15523	15524
880	15919	15909	15900	15894	15889	15887	15886	15886
900	16277	16267	16259	16254	16250	16249	16250	16252
920	16624	16616	16610	16607	16606	16606	16609	16614
940	16959	16953	16950	16950	16952	16957	16963	16971
960	17268	17266	17268	17272	17279	17289	17300	17313
980	17542	17547	17555	17565	17579	17595	17613	17633
1000	17769	17781	17797	17816	17838	17863	17890	17919

TABLE 4
ENTHALPY IN CALORIES PER MOLE

TEMP	PRESSURE IN BARS 4100	4200	4300	4400	4500	4600	4700	4800
20	1805	1837	1870	1902	1935	1967	2000	2032
40	2160	2193	2226	2258	2291	2324	2356	2389
60	2501	2533	2566	2599	2632	2664	2697	2729
80	2830	2863	2896	2928	2961	2993	3026	3058
100	3153	3185	3218	3250	3282	3315	3347	3379
120	3471	3503	3535	3567	3599	3631	3663	3695
140	3791	3823	3854	3886	3918	3949	3981	4012
160	4114	4146	4177	4208	4239	4270	4302	4333
180	4440	4470	4501	4532	4563	4593	4624	4655
200	4766	4796	4826	4857	4887	4917	4948	4978
220	5096	5125	5155	5185	5215	5245	5275	5305
240	5428	5457	5486	5516	5545	5575	5604	5634
260	5761	5789	5818	5847	5876	5905	5934	5963
280	6092	6120	6148	6176	6204	6233	6261	6290
300	6415	6442	6470	6497	6525	6553	6581	6609
320	6741	6768	6795	6822	6849	6876	6903	6931
340	7067	7093	7119	7145	7172	7198	7225	7252
360	7396	7421	7446	7472	7498	7524	7550	7576
380	7740	7765	7789	7814	7839	7864	7889	7915
400	8079	8102	8126	8150	8174	8198	8223	8248
420	8398	8420	8443	8466	8490	8513	8537	8561
440	8716	8738	8760	8782	8804	8827	8850	8874
460	9047	9068	9089	9110	9132	9154	9176	9199
480	9380	9400	9420	9440	9461	9482	9504	9526
500	9724	9743	9762	9782	9802	9822	9843	9864
520	10070	10087	10106	10124	10144	10163	10183	10204
540	10405	10422	10439	10457	10475	10494	10513	10533
560	10741	10756	10773	10790	10808	10826	10844	10863
580	11062	11077	11093	11109	11126	11143	11161	11179
600	11377	11391	11406	11421	11437	11453	11470	11488
620	11686	11699	11713	11727	11742	11758	11774	11791
640	11987	11999	12012	12026	12040	12055	12070	12086
660	12287	12299	12310	12323	12336	12350	12365	12380
680	12584	12594	12605	12616	12629	12642	12656	12670
700	12881	12890	12900	12911	12923	12935	12948	12962
720	13182	13190	13199	13209	13219	13231	13243	13256
740	13491	13498	13506	13514	13524	13534	13546	13558
760	13808	13814	13821	13828	13837	13847	13857	13868
780	14130	14134	14140	14147	14155	14164	14174	14184
800	14464	14468	14473	14480	14487	14495	14504	14514
820	14815	14819	14823	14829	14835	14843	14852	14861
840	15168	15171	15176	15181	15187	15195	15203	15213
860	15525	15528	15533	15538	15545	15553	15561	15571
880	15889	15893	15898	15904	15912	15920	15930	15940
900	16256	16261	16267	16275	16284	16295	16306	16318
920	16620	16628	16637	16647	16659	16672	16685	16700
940	16981	16993	17006	17020	17035	17052	17069	17087
960	17328	17345	17363	17383	17403	17425	17447	17470
980	17655	17679	17704	17731	17758	17787	17816	17845
1000	17951	17983	18018	18053	18089	18126	18164	18202

TABLE 4
ENTHALPY IN CALORIES PER MOLE

TEMP	PRESSURE IN BARS 4900	5000	5100	5200	5300	5400	5500	5600
20	2064	2097	2129	2161	2193	2225	2257	2289
40	2421	2454	2486	2518	2550	2582	2615	2647
60	2762	2794	2826	2859	2891	2923	2955	2987
80	3090	3122	3154	3187	3219	3251	3283	3314
100	3411	3443	3475	3507	3538	3570	3602	3634
120	3727	3758	3790	3822	3853	3885	3916	3948
140	4044	4075	4106	4138	4169	4200	4232	4263
160	4364	4395	4426	4457	4488	4519	4550	4581
180	4686	4716	4747	4778	4809	4839	4870	4901
200	5008	5039	5069	5100	5130	5161	5191	5222
220	5335	5365	5395	5425	5455	5485	5515	5545
240	5663	5693	5723	5752	5782	5812	5842	5871
260	5992	6021	6050	6080	6109	6139	6168	6197
280	6319	6347	6376	6405	6434	6463	6492	6521
300	6637	6665	6694	6722	6751	6779	6808	6837
320	6958	6986	7014	7042	7070	7098	7126	7154
340	7279	7306	7333	7361	7388	7416	7443	7471
360	7602	7629	7655	7682	7709	7736	7763	7791
380	7941	7967	7993	8019	8045	8072	8098	8125
400	8273	8298	8323	8349	8375	8400	8426	8453
420	8586	8610	8635	8660	8685	8710	8736	8761
440	8897	8921	8945	8969	8994	9018	9043	9068
460	9222	9245	9268	9292	9316	9340	9364	9388
480	9548	9570	9593	9616	9639	9662	9686	9710
500	9886	9907	9929	9952	9974	9997	10020	10043
520	10224	10245	10267	10288	10310	10332	10355	10377
540	10553	10573	10594	10615	10636	10658	10680	10702
560	10882	10902	10922	10942	10963	10984	11005	11027
580	11198	11217	11236	11256	11276	11296	11317	11338
600	11506	11524	11543	11562	11581	11601	11621	11642
620	11808	11826	11844	11862	11881	11900	11920	11940
640	12103	12120	12137	12155	12173	12192	12211	12230
660	12396	12412	12429	12446	12464	12482	12500	12519
680	12685	12701	12717	12733	12750	12768	12786	12804
700	12976	12991	13006	13022	13038	13055	13072	13090
720	13269	13283	13298	13313	13328	13345	13361	13378
740	13570	13584	13597	13612	13627	13642	13658	13675
760	13880	13893	13906	13920	13934	13949	13964	13980
780	14195	14207	14220	14233	14246	14261	14275	14291
800	14524	14536	14548	14560	14573	14587	14601	14616
820	14871	14882	14894	14906	14919	14932	14946	14961
840	15223	15233	15245	15257	15270	15284	15298	15312
860	15581	15593	15604	15617	15630	15644	15658	15673
880	15951	15963	15976	15989	16003	16018	16033	16048
900	16330	16344	16358	16373	16388	16404	16420	16437
920	16715	16731	16747	16764	16782	16800	16818	16836
940	17106	17125	17145	17165	17185	17206	17227	17248
960	17494	17518	17542	17566	17591	17616	17640	17665
980	17875	17905	17935	17965	17995	18025	18054	18083
1000	18239	18277	18315	18352	18389	18425	18461	18495

TABLE 4
ENTHALPY IN CALORIES PER MOLE

PRESSURE IN BARS

TEMP	5700	5800	5900	6000	6100	6200	6300	6400
20	2322	2354	2386	2418	2449	2481	2513	2545
40	2679	2711	2743	2775	2807	2839	2871	2902
60	3019	3051	3083	3115	3147	3178	3210	3242
80	3346	3378	3410	3442	3473	3505	3537	3568
100	3665	3697	3729	3760	3792	3823	3855	3886
120	3979	4011	4042	4073	4105	4136	4167	4198
140	4294	4325	4356	4388	4419	4450	4481	4512
160	4612	4643	4674	4705	4736	4767	4797	4828
180	4932	4962	4993	5024	5054	5085	5116	5146
200	5252	5282	5313	5343	5374	5404	5435	5465
220	5575	5606	5636	5666	5696	5726	5756	5787
240	5901	5931	5961	5991	6021	6051	6081	6111
260	6227	6256	6286	6316	6345	6375	6405	6434
280	6550	6580	6609	6638	6667	6697	6726	6755
300	6865	6894	6923	6952	6981	7010	7039	7068
320	7183	7211	7239	7268	7296	7325	7354	7382
340	7499	7527	7555	7583	7611	7639	7668	7696
360	7818	7845	7873	7900	7928	7956	7984	8012
380	8152	8179	8206	8233	8260	8287	8315	8342
400	8479	8505	8532	8558	8585	8612	8639	8666
420	8787	8813	8839	8865	8891	8917	8944	8970
440	9093	9119	9144	9170	9195	9221	9247	9273
460	9413	9438	9463	9488	9513	9538	9564	9589
480	9734	9758	9782	9807	9832	9857	9882	9907
500	10066	10090	10114	10138	10162	10187	10211	10236
520	10400	10423	10447	10470	10494	10518	10542	10566
540	10724	10747	10769	10792	10816	10839	10863	10887
560	11049	11071	11093	11115	11138	11161	11184	11208
580	11359	11381	11403	11425	11447	11469	11492	11515
600	11662	11683	11705	11726	11748	11770	11793	11815
620	11960	11981	12002	12023	12044	12066	12088	12110
640	12250	12270	12291	12311	12332	12353	12375	12396
660	12538	12558	12578	12598	12618	12639	12660	12681
680	12822	12841	12861	12880	12900	12920	12941	12962
700	13108	13126	13145	13164	13183	13203	13223	13244
720	13396	13413	13432	13450	13469	13488	13508	13528
740	13691	13709	13726	13744	13763	13782	13801	13820
760	13996	14013	14030	14048	14065	14084	14102	14121
780	14306	14323	14339	14356	14374	14392	14410	14428
800	14632	14647	14664	14680	14697	14715	14732	14750
820	14976	14992	15007	15024	15040	15058	15075	15093
840	15327	15342	15358	15374	15391	15408	15425	15443
860	15688	15704	15719	15736	15752	15769	15787	15804
880	16064	16080	16096	16113	16130	16147	16165	16183
900	16453	16471	16488	16505	16523	16541	16559	16577
920	16855	16873	16892	16911	16930	16949	16968	16986
940	17269	17290	17310	17331	17352	17372	17392	17412
960	17689	17713	17736	17759	17782	17804	17826	17848
980	18112	18139	18167	18193	18218	18243	18267	18290
1000	18529	18562	18593	18623	18652	18680	18706	18731

TABLE 4
ENTHALPY IN CALORIES PER MOLE

PRESSURE IN BARS

TEMP	6500	6600	6700	6800	6900	7000	7100	7200
20	2577	2609	2641	2672	2704	2736	2768	2799
40	2934	2966	2998	3030	3061	3093	3125	3156
60	3274	3305	3337	3369	3400	3432	3463	3495
80	3600	3631	3663	3694	3726	3757	3789	3820
100	3917	3949	3980	4011	4043	4074	4105	4136
120	4230	4261	4292	4323	4354	4385	4416	4447
140	4543	4574	4605	4636	4667	4698	4729	4759
160	4859	4890	4921	4951	4982	5013	5044	5074
180	5177	5207	5238	5269	5299	5330	5360	5391
200	5495	5526	5556	5586	5617	5647	5678	5708
220	5817	5847	5877	5907	5937	5968	5998	6028
240	6140	6170	6200	6230	6260	6290	6320	6350
260	6464	6494	6523	6553	6583	6612	6642	6672
280	6785	6814	6844	6873	6903	6932	6962	6991
300	7097	7126	7155	7184	7214	7243	7272	7301
320	7411	7440	7469	7498	7527	7556	7585	7614
340	7724	7753	7781	7810	7838	7867	7896	7924
360	8040	8068	8096	8124	8152	8181	8209	8237
380	8370	8397	8425	8453	8481	8509	8537	8565
400	8693	8720	8748	8775	8803	8830	8858	8886
420	8997	9024	9051	9078	9105	9132	9159	9187
440	9300	9326	9352	9379	9406	9433	9459	9486
460	9615	9641	9667	9693	9720	9746	9772	9799
480	9932	9958	9983	10009	10035	10061	10087	10113
500	10261	10286	10311	10336	10362	10387	10413	10439
520	10591	10615	10640	10665	10690	10715	10741	10766
540	10911	10935	10959	10984	11008	11033	11058	11083
560	11231	11255	11279	11303	11328	11352	11377	11402
580	11538	11562	11585	11609	11633	11657	11682	11706
600	11838	11861	11884	11908	11931	11955	11979	12003
620	12132	12155	12177	12200	12224	12247	12271	12295
640	12418	12441	12463	12486	12508	12531	12555	12578
660	12703	12725	12747	12769	12791	12814	12837	12860
680	12983	13004	13026	13048	13070	13092	13115	13137
700	13264	13285	13306	13328	13349	13371	13394	13416
720	13548	13568	13589	13610	13632	13653	13675	13697
740	13840	13860	13880	13901	13922	13943	13964	13986
760	14141	14160	14180	14200	14221	14241	14262	14284
780	14447	14466	14486	14505	14525	14546	14566	14587
800	14769	14788	14807	14826	14846	14866	14886	14907
820	15111	15129	15148	15167	15186	15206	15226	15246
840	15461	15479	15497	15516	15535	15554	15574	15594
860	15822	15840	15858	15877	15896	15915	15934	15953
880	16200	16219	16237	16255	16274	16293	16312	16331
900	16595	16614	16632	16651	16669	16688	16706	16725
920	17005	17024	17043	17061	17080	17098	17116	17134
940	17432	17451	17470	17489	17507	17525	17543	17561
960	17868	17888	17908	17927	17945	17963	17981	17997
980	18312	18333	18353	18373	18391	18408	18425	18440
1000	18755	18777	18798	18817	18835	18851	18866	18880

TABLE 4
ENTHALPY IN CALORIES PER MOLE

TEMP	PRESSURE IN BARS 7300	7400	7500	7600	7700	7800	7900	8000
20	2831	2862	2894	2925	2957	2988	3019	3050
40	3188	3219	3251	3282	3314	3345	3376	3407
60	3526	3558	3589	3621	3652	3683	3714	3746
80	3851	3883	3914	3945	3976	4008	4039	4070
100	4168	4199	4230	4261	4292	4323	4354	4385
120	4478	4509	4540	4571	4602	4633	4664	4695
140	4790	4821	4852	4883	4914	4945	4975	5006
160	5105	5136	5166	5197	5228	5258	5289	5320
180	5421	5452	5482	5513	5543	5574	5604	5635
200	5738	5769	5799	5829	5860	5890	5920	5951
220	6058	6088	6119	6149	6179	6209	6239	6269
240	6380	6410	6440	6470	6500	6530	6560	6590
260	6702	6732	6761	6791	6821	6851	6881	6910
280	7021	7050	7080	7109	7139	7169	7198	7228
300	7331	7360	7389	7419	7448	7478	7507	7536
320	7643	7672	7701	7730	7759	7788	7817	7847
340	7953	7982	8011	8040	8069	8098	8127	8156
360	8266	8294	8323	8351	8380	8409	8438	8466
380	8593	8621	8650	8678	8706	8735	8763	8792
400	8913	8941	8969	8997	9025	9053	9082	9110
420	9214	9242	9269	9297	9325	9353	9381	9409
440	9514	9541	9568	9595	9623	9650	9678	9706
460	9826	9853	9880	9907	9934	9961	9988	10016
480	10140	10166	10193	10219	10246	10273	10300	10327
500	10465	10491	10518	10544	10570	10597	10624	10651
520	10792	10818	10844	10870	10896	10922	10949	10975
540	11109	11134	11160	11186	11212	11238	11264	11290
560	11427	11452	11477	11502	11528	11554	11580	11606
580	11731	11756	11781	11806	11831	11856	11882	11908
600	12027	12052	12077	12101	12126	12152	12177	12202
620	12319	12343	12367	12392	12416	12441	12466	12491
640	12602	12626	12650	12674	12698	12723	12748	12773
660	12883	12907	12931	12955	12979	13003	13027	13052
680	13160	13184	13207	13231	13255	13279	13303	13327
700	13439	13462	13485	13508	13531	13555	13579	13603
720	13719	13742	13765	13788	13811	13834	13858	13882
740	14008	14030	14052	14075	14098	14121	14144	14168
760	14305	14327	14349	14372	14394	14417	14440	14463
780	14609	14630	14652	14674	14696	14718	14741	14764
800	14927	14948	14970	14991	15013	15035	15058	15080
820	15267	15287	15308	15329	15351	15372	15394	15417
840	15614	15634	15655	15675	15696	15717	15739	15761
860	15973	15993	16013	16033	16054	16074	16095	16116
880	16350	16369	16389	16408	16428	16448	16468	16489
900	16744	16763	16781	16800	16819	16838	16857	16877
920	17153	17171	17189	17207	17224	17242	17260	17278
940	17578	17595	17612	17629	17645	17661	17677	17693
960	18014	18029	18044	18059	18073	18087	18100	18113
980	18455	18469	18481	18494	18505	18515	18525	18534
1000	18892	18903	18913	18921	18928	18934	18939	18943

TABLE 4
ENTHALPY IN CALORIES PER MOLE

TEMP	PRESSURE IN BARS							
	8100	8200	8300	8400	8500	8600	8700	8800
20	3081	3112	3143	3174	3204	3235	3265	3296
40	3438	3469	3500	3531	3562	3593	3624	3654
60	3777	3808	3839	3870	3901	3932	3962	3993
80	4101	4132	4163	4194	4225	4256	4287	4318
100	4416	4447	4478	4509	4540	4571	4602	4633
120	4726	4757	4788	4819	4849	4880	4911	4942
140	5037	5068	5098	5129	5160	5190	5221	5252
160	5350	5381	5411	5442	5473	5503	5534	5564
180	5665	5696	5726	5756	5787	5817	5848	5878
200	5981	6011	6041	6072	6102	6132	6163	6193
220	6299	6330	6360	6390	6420	6450	6480	6510
240	6620	6650	6680	6710	6740	6770	6800	6830
260	6940	6970	7000	7030	7059	7089	7119	7149
280	7257	7287	7317	7346	7376	7406	7435	7465
300	7566	7595	7625	7654	7684	7713	7743	7772
320	7876	7905	7934	7964	7993	8022	8052	8081
340	8185	8214	8243	8272	8301	8330	8359	8389
360	8495	8524	8553	8582	8611	8640	8669	8698
380	8820	8849	8878	8906	8935	8964	8993	9022
400	9138	9167	9195	9224	9252	9281	9309	9338
420	9437	9465	9493	9521	9550	9578	9607	9635
440	9734	9762	9789	9818	9846	9874	9902	9930
460	10043	10071	10099	10127	10154	10182	10210	10239
480	10355	10382	10410	10437	10465	10492	10520	10548
500	10678	10705	10732	10759	10787	10814	10842	10869
520	11002	11029	11056	11083	11110	11137	11165	11192
540	11316	11343	11370	11396	11423	11450	11477	11505
560	11632	11658	11684	11711	11738	11764	11791	11818
580	11934	11960	11986	12012	12038	12065	12091	12118
600	12228	12254	12279	12305	12332	12358	12384	12411
620	12517	12542	12568	12593	12619	12645	12671	12698
640	12798	12823	12848	12874	12899	12925	12951	12977
660	13077	13102	13127	13152	13177	13203	13228	13254
680	13352	13376	13401	13426	13451	13476	13502	13527
700	13627	13652	13676	13701	13726	13751	13776	13801
720	13906	13930	13954	13978	14003	14028	14053	14078
740	14192	14215	14239	14264	14288	14313	14337	14362
760	14486	14510	14534	14558	14582	14606	14631	14655
780	14787	14810	14834	14857	14881	14905	14930	14954
800	15103	15126	15149	15172	15196	15220	15244	15268
820	15439	15461	15484	15507	15530	15554	15577	15601
840	15782	15805	15827	15849	15872	15895	15918	15942
860	16138	16159	16181	16203	16225	16247	16269	16292
880	16509	16530	16550	16571	16593	16614	16635	16657
900	16896	16915	16935	16955	16974	16994	17015	17035
920	17296	17313	17331	17349	17367	17385	17403	17421
940	17708	17724	17739	17755	17770	17785	17800	17816
960	18126	18138	18150	18162	18174	18185	18196	18207
980	18543	18551	18558	18565	18571	18577	18583	18588
1000	18946	18948	18949	18949	18948	18947	18944	18942

TABLE 4
ENTHALPY IN CALORIES PER MOLE

TEMP	PRESSURE IN BARS 8900	9000	9100	9200	9300	9400	9500	9600
20	(3326)	(3356)	(3386)	(3416)	(3446)	(3475)	(3505)	(3534)
40	3685	3715	3745	3776	3806	3836	3866	3896
60	4024	4054	4085	4116	4146	4176	4207	4237
80	4348	4379	4410	4440	4471	4502	4532	4563
100	4663	4694	4725	4756	4786	4817	4848	4878
120	4972	5003	5034	5064	5095	5126	5156	5187
140	5282	5313	5343	5374	5405	5435	5466	5496
160	5595	5625	5656	5686	5717	5747	5778	5808
180	5908	5939	5969	6000	6030	6060	6091	6121
200	6223	6253	6284	6314	6344	6374	6404	6435
220	6540	6571	6601	6631	6661	6691	6721	6751
240	6860	6890	6920	6950	6980	7010	7040	7070
260	7179	7209	7238	7268	7298	7328	7358	7387
280	7495	7524	7554	7584	7614	7643	7673	7703
300	7802	7831	7861	7891	7920	7950	7979	8009
320	8111	8140	8169	8199	8228	8258	8287	8317
340	8418	8447	8476	8506	8535	8564	8594	8623
360	8727	8756	8785	8814	8844	8873	8902	8932
380	9051	9079	9109	9138	9167	9196	9225	9254
400	9367	9396	9425	9454	9482	9512	9541	9570
420	9664	9692	9721	9750	9779	9808	9837	9866
440	9959	9987	10016	10045	10073	10102	10131	10160
460	10267	10295	10323	10352	10380	10409	10438	10467
480	10576	10604	10632	10661	10689	10718	10746	10775
500	10897	10925	10953	10981	11009	11038	11066	11094
520	11220	11247	11275	11303	11331	11359	11387	11415
540	11532	11559	11587	11615	11642	11670	11698	11726
560	11845	11872	11900	11927	11955	11982	12010	12038
580	12145	12172	12199	12226	12253	12281	12308	12336
600	12437	12464	12491	12517	12544	12572	12599	12626
620	12724	12750	12777	12804	12830	12857	12884	12911
640	13003	13029	13055	13082	13108	13135	13161	13188
660	13280	13306	13332	13358	13384	13411	13437	13464
680	13553	13578	13604	13630	13656	13682	13708	13735
700	13827	13852	13878	13903	13929	13955	13981	14007
720	14103	14128	14154	14179	14205	14230	14256	14282
740	14387	14412	14437	14463	14488	14514	14539	14565
760	14680	14705	14730	14755	14780	14806	14831	14857
780	14979	15003	15028	15053	15078	15103	15129	15154
800	15292	15317	15341	15366	15391	15416	15441	15466
820	15625	15649	15674	15698	15723	15748	15773	15798
840	15965	15989	16013	16037	16061	16085	16110	16135
860	16315	16338	16361	16385	16408	16432	16456	16480
880	16679	16701	16723	16746	16768	16791	16815	16838
900	17055	17076	17097	17118	17139	17160	17182	17204
920	17440	17458	17477	17496	17515	17534	17553	17573
940	17831	17846	17862	17877	17893	17909	17925	17942
960	18218	18229	18240	18251	18262	18274	18285	18296
980	18593	18598	18602	18607	18612	18616	18621	18625
1000	18939	18935	18931	18927	18922	18917	18913	18908

TABLE 4
ENTHALPY IN CALORIES PER MOLE

TEMP	PRESSURE IN BARS 9700	9800	9900	10000
20	(3563)	(3592)	(3622)	(3651)
40	3926	3955	3985	4015
60	4267	4297	4328	4358
80	4593	4623	4654	4684
100	4908	4939	4970	5000
120	5217	5248	5278	5309
140	5526	5557	5588	5618
160	5838	5869	5899	5930
180	6151	6181	6212	6242
200	6464	6495	6525	6555
220	6781	6811	6841	6871
240	7099	7129	7159	7189
260	7417	7447	7476	7506
280	7732	7762	7791	7821
300	8038	8068	8097	8127
320	8346	8375	8405	8434
340	8652	8682	8711	8741
360	8960	8990	9019	9049
380	9283	9312	9342	9371
400	9598	9628	9657	9686
420	9894	9923	9953	9982
440	10188	10217	10246	10276
460	10495	10524	10553	10582
480	10803	10832	10861	10890
500	11123	11151	11180	11209
520	11443	11472	11500	11529
540	11754	11782	11810	11839
560	12065	12093	12121	12150
580	12363	12391	12419	12447
600	12653	12681	12708	12736
620	12938	12965	12992	13020
640	13215	13242	13269	13296
660	13490	13516	13543	13570
680	13760	13787	13813	13840
700	14033	14059	14085	14111
720	14307	14333	14359	14386
740	14590	14616	14642	14668
760	14882	14908	14933	14959
780	15179	15205	15231	15256
800	15491	15517	15543	15568
820	15822	15848	15873	15899
840	16159	16184	16210	16235
860	16504	16529	16554	16579
880	16861	16885	16909	16933
900	17226	17248	17271	17294
920	17592	17613	17633	17654
940	17958	17975	17992	18010
960	18308	18320	18332	18345
980	18630	18635	18640	18646
1000	18903	18898	18894	18890

TABLE 5
FUGACITY IN BARS

TEMP	PRESSURE IN BARS 100	200	300	400	500	600	700	800
20	0	0	0	0	0	0	0	0
40	0	0	0	0	0	0	0	0
60	0	0	0	0	0	0	0	0
80	0	1	1	1	1	1	1	1
100	1	1	1	1	1	1	2	2
120	2	2	2	2	3	3	3	3
140	4	4	4	4	5	5	5	5
160	6	7	7	7	8	8	9	9
180	10	10	11	12	12	13	13	14
200	15	16	16	17	18	19	20	21
220	22	23	24	25	26	28	29	31
240	30	32	33	35	37	39	41	43
260	41	43	45	47	50	52	55	57
280	53	56	59	62	65	68	71	75
300	68	71	75	79	83	87	91	96
320	78	89	93	98	103	108	114	119
340	81	108	114	120	126	132	139	145
360	83	128	135	143	150	158	166	174
380	85	140	158	168	177	186	195	205
400	87	147	179	192	204	215	227	238
420	88	153	195	216	231	245	259	272
440	89	158	208	238	258	275	291	307
460	91	163	217	256	282	303	323	341
480	92	166	225	270	304	330	353	375
500	92	170	233	284	323	355	383	409
520	93	173	240	295	340	378	411	440
540	94	175	245	305	355	398	435	469
560	94	178	250	314	369	417	459	498
580	95	179	254	321	381	432	479	522
600	95	181	259	328	390	446	498	546
620	96	183	262	334	401	460	516	566
640	96	184	266	341	410	473	532	586
660	96	186	268	345	416	483	545	605
680	96	187	271	351	424	494	559	621
700	97	188	274	355	431	503	571	637
720	97	189	276	359	437	511	583	652
740	97	190	278	362	443	519	594	663
760	98	191	280	365	448	526	603	677
780	98	192	281	368	452	531	610	686
800	98	193	283	371	456	539	619	697

TABLE 5
FUGACITY IN BARS

TEMP	PRESSURE IN BARS 900	1000	1100	1200	1300	1400	1500	1600
20	0	0	0	0	0	0	0	0
40	0	0	0	0	0	0	0	0
60	0	0	0	0	0	0	1	1
80	1	1	1	1	1	1	1	1
100	2	2	2	2	2	2	2	3
120	3	3	4	4	4	4	4	5
140	6	6	6	7	7	7	8	8
160	9	10	11	11	12	12	13	14
180	15	16	17	17	18	19	20	21
200	22	23	25	26	27	29	30	31
220	32	34	35	37	39	41	43	45
240	45	47	49	52	54	57	59	62
260	60	63	66	69	73	76	79	83
280	78	82	86	90	94	99	103	108
300	100	105	110	115	120	126	131	137
320	125	131	137	143	150	156	163	171
340	152	160	167	175	182	191	199	208
360	183	191	200	209	219	228	238	249
380	215	225	236	247	258	269	281	293
400	250	262	274	287	300	314	327	342
420	285	300	315	329	344	360	376	392
440	323	339	355	372	389	407	425	444
460	360	379	398	417	436	456	477	498
480	396	418	440	461	483	506	529	552
500	434	459	483	508	532	558	583	609
520	468	497	525	552	580	608	636	665
540	502	532	563	593	624	655	686	718
560	533	569	603	637	671	705	740	775
580	562	601	639	676	713	750	787	825
600	590	631	673	713	754	794	834	875
620	617	661	706	751	794	838	882	926
640	640	690	738	786	833	881	928	975
660	660	715	768	819	870	920	970	1021
680	681	738	794	849	903	957	1010	1064
700	699	761	820	879	936	993	1050	1107
720	718	781	843	905	965	1026	1086	1146
740	733	799	865	929	993	1056	1120	1183
760	747	819	887	955	1022	1088	1155	1221
780	760	831	902	972	1041	1110	1179	1248
800	773	848	921	994	1067	1139	1211	1283
820		861	937	1012	1087	1162	1236	1311
840		873	952	1029	1107	1184	1261	1338
860		884	965	1045	1124	1203	1283	1363
880		895	977	1059	1141	1222	1304	1386
900		904	989	1073	1157	1240	1324	1408
920		913	999	1084	1170	1255	1340	1426
940		921	1008	1095	1182	1269	1356	1444
960		928	1016	1105	1193	1281	1370	1459
980		934	1024	1113	1203	1293	1383	1473
1000		940	1031	1121	1212	1303	1395	1487

TABLE 5
FUGACITY IN BARS

TEMP	PRESSURE IN BARS 1700	1800	1900	2000	2100	2200	2300	2400
20	0	0	0	0	0	0	0	0
40	0	0	0	0	0	0	0	0
60	1	1	1	1	1	1	1	1
80	1	1	1	2	2	2	2	2
100	3	3	3	3	3	4	4	4
120	5	5	6	6	6	7	7	7
140	9	9	10	10	11	11	12	13
160	14	15	16	17	18	18	19	20
180	22	23	25	26	27	29	30	31
200	33	35	36	38	40	42	44	46
220	47	50	52	54	57	60	62	65
240	65	68	71	75	78	82	85	89
260	87	91	95	100	104	109	114	119
280	113	118	123	129	135	141	147	153
300	143	150	156	163	170	178	185	193
320	178	186	194	202	211	220	230	239
340	217	226	236	246	257	268	279	290
360	260	271	282	294	307	319	333	346
380	306	319	333	347	361	376	391	407
400	356	371	387	403	420	437	455	473
420	409	426	444	463	482	501	522	543
440	463	483	503	524	545	568	590	614
460	519	541	564	588	612	637	662	689
480	576	601	626	652	679	707	735	764
500	636	664	692	721	751	781	812	845
520	695	725	756	788	820	854	888	923
540	750	784	817	852	886	923	960	998
560	810	846	883	921	958	997	1037	1079
580	864	903	943	983	1022	1064	1108	1152
600	917	959	1002	1046	1087	1132	1178	1226
620	971	1016	1062	1109	1152	1201	1251	1301
640	1023	1072	1121	1171	1217	1269	1322	1376
660	1073	1124	1177	1230	1279	1334	1390	1447
680	1119	1174	1229	1286	1338	1396	1455	1515
700	1165	1223	1282	1342	1397	1458	1521	1584
720	1207	1268	1330	1393	1452	1516	1582	1648
740	1247	1311	1376	1442	1505	1572	1641	1711
760	1288	1356	1424	1493	1560	1631	1703	1775
780	1318	1388	1458	1530	1601	1675	1749	1824
800	1355	1428	1502	1576	1652	1728	1805	1884
820	1386	1462	1538	1615	1694	1773	1853	1935
840	1416	1494	1573	1652	1734	1816	1899	1983
860	1443	1523	1605	1687	1771	1855	1940	2027
880	1468	1551	1635	1720	1805	1892	1979	2068
900	1492	1577	1663	1750	1837	1926	2016	2107
920	1512	1599	1687	1776	1865	1956	2048	2141
940	1532	1620	1710	1801	1892	1985	2079	2174
960	1548	1639	1730	1823	1916	2011	2106	2203
980	1565	1657	1750	1844	1939	2035	2133	2232
1000	1580	1673	1768	1864	1961	2059	2158	2259

TABLE 5
FUGACITY IN BARS

TEMP	PRESSURE IN BARS 2500	2600	2700	2800	2900	3000	3100	3200
20	0	0	0	0	0	0	0	0
40	0	0	0	0	1	1	1	1
60	1	1	1	1	1	1	1	1
80	2	2	2	3	3	3	3	3
100	4	4	5	5	5	6	6	6
120	8	8	9	9	10	10	11	11
140	13	14	15	15	16	17	18	19
160	21	22	24	25	26	27	29	30
180	33	35	36	38	40	42	44	46
200	48	51	53	55	58	61	64	67
220	68	72	75	78	82	86	89	94
240	93	98	102	107	111	116	121	127
260	124	129	135	141	147	154	160	167
280	160	167	174	181	189	197	206	214
300	202	210	219	228	238	248	258	269
320	249	260	271	282	293	305	318	331
340	302	315	328	341	355	369	384	399
360	360	375	390	406	422	439	456	474
380	424	441	458	477	495	515	535	556
400	492	512	532	553	574	597	620	643
420	564	586	609	633	658	683	709	736
440	638	663	689	716	743	771	800	830
460	716	744	773	802	833	864	896	930
480	794	825	857	890	923	958	993	1030
500	878	912	947	983	1019	1057	1097	1137
520	959	996	1034	1074	1114	1155	1198	1241
540	1037	1077	1118	1161	1204	1249	1294	1341
560	1121	1164	1209	1255	1302	1350	1399	1450
580	1197	1244	1291	1340	1390	1442	1494	1548
600	1274	1324	1375	1427	1480	1535	1591	1648
620	1353	1406	1460	1515	1572	1630	1689	1750
640	1431	1487	1544	1603	1663	1724	1787	1852
660	1505	1564	1625	1687	1750	1815	1881	1949
680	1576	1639	1703	1768	1835	1903	1972	2043
700	1649	1715	1782	1850	1920	1992	2065	2139
720	1716	1785	1855	1927	2000	2074	2151	2228
740	1781	1853	1927	2002	2078	2156	2235	2316
760	1850	1925	2002	2080	2159	2240	2323	2407
780	1901	1979	2058	2139	2221	2305	2390	2477
800	1964	2045	2127	2211	2297	2383	2472	2562
820	2017	2101	2186	2273	2361	2451	2542	2636
840	2068	2155	2242	2332	2423	2516	2610	2706
860	2114	2204	2294	2386	2480	2575	2672	2770
880	2158	2250	2343	2437	2533	2631	2730	2831
900	2199	2293	2388	2485	2584	2684	2785	2889
920	2236	2332	2429	2528	2629	2731	2835	2940
940	2271	2369	2468	2569	2672	2776	2882	2990
960	2302	2402	2503	2606	2710	2816	2924	3034
980	2332	2434	2537	2642	2748	2856	2965	3076
1000	2361	2464	2569	2675	2783	2892	3003	3115

TABLE 5
FUGACITY IN BARS

PRESSURE IN BARS

TEMP	3300	3400	3500	3600	3700	3800	3900	4000
20	0	0	0	0	0	0	0	0
40	1	1	1	1	1	1	1	1
60	2	2	2	2	2	2	2	2
80	3	4	4	4	4	4	5	5
100	7	7	7	8	8	9	9	10
120	12	12	13	14	14	15	16	17
140	20	21	22	23	24	25	27	28
160	32	33	35	37	38	40	42	44
180	48	50	53	55	58	61	64	67
200	70	73	76	80	84	87	91	96
220	98	102	107	112	117	122	127	133
240	132	138	144	151	157	164	171	178
260	174	182	190	198	206	215	224	233
280	223	233	242	252	263	274	285	297
300	280	291	303	315	328	342	355	369
320	344	358	372	387	402	418	435	452
340	415	432	449	466	484	503	523	543
360	493	512	532	552	573	595	618	641
380	577	599	622	645	670	695	721	748
400	668	693	719	746	774	802	832	863
420	763	792	821	851	883	915	948	982
440	861	893	926	960	994	1030	1067	1105
460	964	999	1035	1073	1111	1151	1192	1234
480	1068	1106	1146	1187	1230	1273	1318	1364
500	1178	1220	1264	1309	1355	1403	1452	1502
520	1286	1332	1380	1428	1478	1530	1583	1637
540	1390	1439	1490	1542	1596	1651	1708	1766
560	1501	1555	1609	1666	1723	1782	1843	1906
580	1603	1660	1718	1778	1839	1902	1966	2033
600	1707	1767	1828	1892	1957	2023	2091	2161
620	1812	1876	1942	2009	2077	2148	2220	2293
640	1917	1985	2054	2124	2197	2271	2347	2425
660	2018	2089	2162	2236	2312	2390	2470	2551
680	2116	2190	2266	2344	2423	2505	2588	2673
700	2215	2293	2373	2454	2537	2622	2709	2798
720	2308	2389	2472	2556	2643	2731	2822	2914
740	2398	2483	2569	2657	2747	2839	2933	3029
760	2493	2581	2671	2762	2856	2951	3049	3149
780	2566	2657	2749	2843	2940	3038	3138	3241
800	2654	2748	2844	2942	3041	3143	3247	3353
820	2731	2827	2926	3027	3129	3234	3341	3450
840	2804	2903	3005	3108	3214	3321	3431	3543
860	2871	2973	3077	3183	3291	3402	3514	3629
880	2934	3039	3146	3254	3365	3478	3593	3710
900	2994	3101	3210	3322	3435	3550	3667	3787
920	3048	3157	3268	3382	3497	3614	3733	3855
940	3099	3210	3324	3439	3556	3675	3797	3920
960	3145	3258	3373	3489	3608	3729	3852	3977
980	3189	3303	3420	3538	3658	3780	3905	4031
1000	3230	3345	3463	3582	3704	3827	3952	4079

TABLE 5
FUGACITY IN BARS

TEMP	PRESSURE IN BARS 4100	4200	4300	4400	4500	4600	4700	4800
20	0	0	0	0	1	1	1	1
40	1	1	1	1	1	1	2	2
60	3	3	3	3	3	3	4	4
80	5	6	6	6	7	7	7	8
100	10	11	11	12	12	13	14	15
120	18	19	20	21	22	23	24	25
140	30	31	33	34	36	38	40	42
160	46	49	51	54	56	59	62	65
180	70	73	76	80	84	88	92	96
200	100	104	109	114	119	125	130	136
220	139	145	151	158	165	172	179	187
240	186	194	202	211	220	229	239	249
260	243	253	264	274	286	298	310	323
280	309	321	334	348	362	376	392	407
300	384	400	415	432	449	467	485	504
320	470	488	507	526	547	568	590	612
340	564	585	607	630	654	679	704	731
360	665	690	716	743	770	799	828	859
380	776	804	834	864	896	928	962	997
400	894	927	960	995	1030	1067	1105	1144
420	1018	1054	1092	1131	1171	1212	1254	1298
440	1145	1185	1227	1270	1314	1360	1407	1455
460	1277	1322	1368	1415	1464	1514	1566	1619
480	1411	1460	1510	1562	1615	1670	1726	1784
500	1554	1607	1661	1718	1776	1835	1896	1959
520	1693	1750	1809	1870	1932	1996	2062	2130
540	1826	1887	1950	2015	2081	2150	2220	2292
560	1970	2035	2103	2172	2243	2316	2391	2468
580	2100	2170	2241	2315	2390	2467	2547	2628
600	2233	2307	2382	2459	2539	2620	2704	2789
620	2369	2447	2526	2608	2692	2777	2865	2956
640	2504	2586	2670	2756	2843	2933	3026	3120
660	2635	2720	2808	2898	2989	3084	3180	3279
680	2760	2850	2941	3035	3130	3229	3329	3432
700	2889	2982	3077	3175	3275	3377	3481	3588
720	3009	3106	3204	3306	3409	3515	3623	3734
740	3127	3227	3330	3434	3542	3651	3763	3878
760	3250	3354	3461	3569	3681	3794	3910	4029
780	3345	3452	3562	3673	3787	3904	4023	4144
800	3461	3572	3684	3800	3917	4037	4160	4286
820	3561	3675	3790	3909	4030	4153	4279	4408
840	3657	3774	3892	4014	4138	4264	4393	4525
860	3746	3865	3987	4111	4237	4366	4498	4633
880	3830	3951	4076	4202	4332	4463	4598	4735
900	3909	4033	4159	4288	4420	4554	4691	4831
920	3979	4105	4234	4365	4499	4635	4774	4915
940	4046	4174	4304	4437	4573	4711	4851	4994
960	4104	4234	4366	4500	4636	4776	4918	5062
980	4159	4290	4423	4558	4696	4836	4979	5125
1000	4209	4340	4474	4610	4748	4889	5032	5178

TABLE 5
FUGACITY IN BARS

TEMP	PRESSURE IN BARS 4900	5000	5100	5200	5300	5400	5500	5600
20	1	1	1	1	1	1	1	1
40	2	2	2	2	2	2	3	3
60	4	4	4	5	5	5	6	6
80	8	9	9	10	10	11	11	12
100	15	16	17	18	19	20	21	22
120	27	28	29	31	33	34	36	38
140	44	46	48	51	53	56	58	61
160	68	71	74	78	82	85	89	94
180	100	105	110	115	120	126	131	137
200	142	148	155	162	169	177	184	192
220	195	204	212	221	231	241	251	262
240	259	270	282	293	306	318	332	345
260	336	349	364	379	394	410	426	443
280	423	440	458	476	495	514	534	555
300	523	544	565	587	609	633	657	682
320	635	660	685	710	737	765	794	823
340	758	786	816	846	877	909	942	977
360	890	923	956	991	1027	1064	1102	1141
380	1033	1070	1108	1147	1188	1230	1273	1318
400	1185	1227	1270	1314	1360	1407	1456	1506
420	1343	1390	1438	1488	1539	1591	1645	1701
440	1505	1557	1610	1664	1720	1778	1838	1899
460	1674	1730	1789	1848	1910	1973	2038	2106
480	1844	1905	1968	2033	2100	2169	2239	2312
500	2024	2090	2159	2229	2301	2376	2452	2531
520	2199	2271	2344	2420	2498	2577	2659	2743
540	2366	2442	2521	2601	2684	2768	2855	2945
560	2547	2629	2712	2798	2885	2976	3068	3163
580	2711	2797	2885	2975	3068	3162	3260	3360
600	2877	2967	3060	3155	3252	3352	3454	3559
620	3048	3143	3240	3339	3441	3546	3653	3763
640	3217	3316	3418	3523	3629	3739	3851	3966
660	3380	3484	3590	3698	3810	3924	4041	4160
680	3537	3645	3755	3868	3984	4102	4224	4348
700	3698	3810	3925	4042	4162	4285	4411	4540
720	3848	3964	4082	4204	4328	4455	4585	4718
740	3995	4115	4238	4363	4491	4622	4757	4894
760	4150	4274	4401	4530	4663	4799	4937	5079
780	4269	4396	4526	4658	4794	4933	5074	5219
800	4414	4545	4678	4815	4955	5097	5243	5392
820	4539	4673	4810	4950	5093	5239	5388	5541
840	4659	4797	4937	5080	5226	5375	5528	5683
860	4770	4910	5053	5199	5348	5500	5656	5814
880	4875	5018	5163	5312	5464	5618	5776	5938
900	4973	5118	5266	5417	5571	5728	5889	6052
920	5059	5206	5357	5510	5666	5825	5987	6153
940	5140	5289	5441	5596	5753	5914	6078	6246
960	5209	5359	5512	5668	5827	5989	6154	6323
980	5273	5424	5578	5734	5894	6057	6223	6392
1000	5327	5478	5632	5789	5949	6112	6279	6448

TABLE 5
FUGACITY IN BARS

PRESSURE IN BARS

TEMP	5700	5800	5900	6000	6100	6200	6300	6400
20	1	1	1	1	1	2	2	2
40	3	3	3	3	4	4	4	4
60	6	7	7	7	8	8	9	9
80	13	13	14	15	16	17	18	18
100	23	25	26	27	29	30	32	33
120	40	42	44	46	48	51	53	56
140	64	67	71	74	78	81	85	89
160	98	103	107	112	118	123	129	135
180	143	150	157	164	171	179	187	195
200	201	210	219	228	238	249	259	271
220	273	285	297	309	322	336	350	365
240	360	374	390	406	422	440	458	476
260	461	480	499	519	539	561	583	606
280	577	600	623	647	672	698	725	753
300	708	735	763	792	822	853	885	918
320	854	886	919	953	988	1024	1062	1101
340	1013	1049	1087	1127	1167	1210	1253	1298
360	1182	1224	1268	1313	1359	1407	1456	1507
380	1364	1412	1461	1512	1564	1618	1674	1732
400	1558	1611	1667	1723	1782	1843	1905	1969
420	1759	1818	1879	1942	2007	2074	2143	2214
440	1963	2028	2095	2164	2235	2308	2384	2461
460	2175	2246	2319	2394	2472	2551	2633	2718
480	2387	2464	2543	2624	2708	2794	2883	2974
500	2612	2695	2780	2868	2958	3051	3146	3244
520	2830	2919	3010	3104	3200	3300	3401	3506
540	3036	3131	3228	3327	3429	3534	3641	3752
560	3261	3361	3464	3569	3677	3788	3902	4019
580	3462	3568	3675	3786	3900	4016	4136	4258
600	3667	3777	3890	4006	4125	4247	4372	4500
620	3876	3992	4110	4231	4356	4483	4614	4748
640	4084	4204	4328	4455	4585	4718	4854	4993
660	4283	4409	4537	4669	4804	4942	5083	5228
680	4475	4605	4738	4875	5015	5158	5304	5454
700	4672	4807	4945	5086	5231	5379	5530	5685
720	4854	4994	5136	5282	5431	5583	5739	5899
740	5034	5178	5324	5474	5628	5785	5945	6109
760	5223	5371	5523	5677	5836	5997	6162	6331
780	5367	5518	5673	5831	5992	6157	6326	6498
800	5544	5700	5858	6021	6186	6356	6528	6705
820	5696	5855	6017	6183	6352	6525	6702	6882
840	5842	6004	6170	6339	6512	6688	6868	7052
860	5976	6141	6310	6482	6658	6837	7020	7207
880	6102	6270	6441	6616	6795	6977	7162	7352
900	6219	6390	6563	6741	6922	7106	7294	7486
920	6322	6494	6670	6849	7032	7218	7408	7603
940	6416	6590	6768	6948	7133	7321	7513	7709
960	6495	6670	6848	7030	7216	7405	7598	7795
980	6564	6740	6920	7102	7289	7479	7673	7870
1000	6621	6797	6976	7159	7346	7536	7730	7928

TABLE 5
FUGACITY IN BARS

TEMP	PRESSURE IN BARS							
	6500	6600	6700	6800	6900	7000	7100	7200
20	2	2	2	2	2	3	3	3
40	5	5	5	5	6	6	6	7
60	10	11	11	12	12	13	14	15
80	19	21	22	23	24	25	27	28
100	35	37	39	41	43	45	48	50
120	59	62	65	68	72	75	79	83
140	94	98	103	108	113	118	124	130
160	141	148	155	162	169	177	185	193
180	204	213	223	232	243	254	265	276
200	282	295	307	320	334	348	363	379
220	380	396	412	429	447	466	485	505
240	495	516	536	558	580	604	628	653
260	630	655	681	707	735	764	794	824
280	782	812	843	875	909	943	979	1016
300	952	988	1025	1063	1102	1143	1186	1229
320	1141	1182	1226	1270	1316	1364	1413	1464
340	1344	1392	1442	1493	1546	1600	1657	1715
360	1560	1615	1671	1729	1789	1851	1914	1980
380	1791	1852	1915	1981	2048	2117	2189	2262
400	2035	2104	2174	2247	2321	2398	2478	2560
420	2287	2362	2440	2520	2602	2687	2774	2864
440	2541	2624	2708	2796	2885	2978	3073	3171
460	2805	2894	2986	3081	3178	3278	3381	3487
480	3067	3164	3263	3364	3469	3576	3687	3801
500	3345	3448	3554	3664	3776	3891	4010	4131
520	3613	3723	3836	3953	4072	4195	4321	4450
540	3865	3982	4101	4224	4350	4479	4611	4748
560	4139	4262	4389	4518	4651	4788	4928	5071
580	4384	4513	4645	4781	4920	5062	5208	5358
600	4631	4766	4904	5046	5191	5339	5492	5648
620	4885	5026	5170	5317	5469	5623	5782	5945
640	5136	5282	5432	5586	5743	5904	6069	6238
660	5376	5528	5683	5842	6005	6172	6342	6517
680	5607	5764	5924	6089	6257	6429	6605	6785
700	5843	6005	6171	6341	6514	6692	6873	7059
720	6062	6229	6399	6574	6752	6934	7121	7311
740	6277	6448	6624	6803	6986	7173	7364	7560
760	6504	6680	6860	7044	7233	7425	7621	7822
780	6674	6854	7037	7225	7416	7612	7812	8016
800	6885	7070	7258	7450	7646	7847	8051	8260
820	7066	7254	7446	7642	7842	8046	8254	8467
840	7239	7430	7626	7825	8029	8236	8448	8664
860	7397	7592	7790	7992	8199	8410	8625	8844
880	7545	7743	7944	8149	8359	8572	8790	9012
900	7682	7882	8086	8294	8506	8722	8942	9167
920	7800	8002	8208	8418	8632	8850	9073	9299
940	7908	8112	8319	8531	8747	8967	9191	9420
960	7996	8200	8409	8622	8839	9060	9286	9516
980	8072	8277	8487	8701	8919	9141	9367	9598
1000	8130	8336	8546	8760	8978	9201	9428	9660

TABLE 5
FUGACITY IN BARS

TEMP	PRESSURE IN BARS 7300	7400	7500	7600	7700	7800	7900	8000
20	3	3	3	4	4	4	4	5
40	7	8	8	9	9	10	10	11
60	16	16	17	18	19	20	22	23
80	30	31	33	35	37	39	41	43
100	53	55	58	61	64	68	71	75
120	87	91	96	101	105	111	116	122
140	136	143	149	156	164	172	180	188
160	202	212	221	231	242	253	264	276
180	289	301	315	328	343	358	373	390
200	395	412	429	447	466	486	507	528
220	526	547	570	593	618	643	669	696
240	679	706	735	764	794	826	858	892
260	856	890	924	960	997	1035	1075	1116
280	1054	1094	1135	1178	1222	1268	1315	1364
300	1275	1321	1370	1420	1472	1526	1581	1639
320	1516	1571	1627	1685	1745	1807	1871	1938
340	1775	1837	1902	1968	2036	2107	2180	2256
360	2048	2118	2191	2266	2343	2422	2504	2589
380	2338	2417	2498	2581	2667	2756	2847	2941
400	2644	2731	2820	2913	3008	3106	3207	3311
420	2957	3052	3151	3252	3356	3463	3574	3687
440	3272	3375	3482	3592	3705	3821	3940	4063
460	3596	3708	3823	3941	4063	4188	4317	4450
480	3917	4038	4161	4288	4418	4552	4690	4831
500	4256	4385	4517	4652	4792	4935	5081	5232
520	4583	4719	4859	5002	5150	5301	5457	5616
540	4887	5031	5178	5328	5483	5642	5805	5972
560	5218	5369	5524	5683	5846	6013	6184	6360
580	5512	5669	5831	5996	6166	6340	6518	6701
600	5808	5972	6140	6312	6488	6669	6854	7044
620	6111	6282	6456	6635	6819	7006	7198	7395
640	6411	6587	6769	6954	7144	7338	7537	7741
660	6696	6879	7066	7257	7453	7654	7859	8069
680	6969	7157	7350	7547	7749	7955	8166	8382
700	7249	7443	7641	7844	8052	8264	8481	8703
720	7506	7706	7909	8117	8330	8548	8770	8997
740	7760	7964	8172	8386	8603	8826	9053	9286
760	8027	8237	8451	8669	8892	9121	9353	9591
780	8225	8438	8655	8877	9104	9335	9572	9813
800	8473	8691	8913	9140	9372	9608	9850	10096
820	8684	8905	9131	9362	9598	9838	10083	10333
840	8885	9110	9340	9574	9813	10057	10306	10560
860	9068	9296	9529	9767	10009	10256	10508	10765
880	9239	9470	9706	9946	10191	10441	10696	10956
900	9396	9630	9868	10111	10359	10611	10869	11131
920	9531	9766	10007	10252	10501	10756	11015	11280
940	9653	9890	10133	10379	10631	10887	11148	11415
960	9750	9989	10232	10480	10733	10991	11254	11521
980	9833	10073	10318	10567	10821	11080	11344	11613
1000	9896	10136	10382	10632	10887	11147	11412	11682

TABLE 5
FUGACITY IN BARS

PRESSURE IN BARS

TEMP	8100	8200	8300	8400	8500	8600	8700	8800
20	5	5	6	6	6	7	7	8
40	12	12	13	14	15	16	16	17
60	24	26	27	28	30	32	34	35
80	45	48	50	53	56	59	62	65
100	79	83	87	91	96	101	106	112
120	128	134	141	147	155	162	170	178
140	197	206	216	226	237	248	259	271
160	289	302	315	329	344	360	376	392
180	407	424	443	462	482	503	524	547
200	550	573	597	622	648	675	703	733
220	725	754	785	816	849	884	919	956
240	927	964	1002	1041	1082	1124	1168	1214
260	1159	1203	1249	1297	1346	1397	1450	1505
280	1415	1468	1522	1579	1637	1698	1760	1825
300	1698	1759	1823	1889	1957	2027	2100	2175
320	2006	2077	2150	2226	2304	2384	2468	2554
340	2334	2414	2497	2583	2671	2763	2857	2954
360	2676	2767	2860	2956	3054	3157	3262	3370
380	3039	3139	3242	3348	3458	3571	3688	3808
400	3418	3528	3642	3759	3880	4004	4132	4264
420	3804	3925	4049	4176	4308	4443	4582	4725
440	4190	4320	4454	4592	4733	4879	5029	5183
460	4586	4726	4870	5018	5170	5326	5486	5651
480	4976	5126	5279	5437	5599	5765	5936	6111
500	5387	5546	5709	5877	6049	6226	6407	6593
520	5780	5948	6120	6297	6479	6665	6856	7052
540	6144	6320	6500	6685	6875	7070	7269	7474
560	6540	6725	6914	7108	7307	7511	7720	7934
580	6888	7080	7276	7478	7684	7895	8112	8334
600	7238	7437	7641	7850	8063	8282	8506	8736
620	7597	7803	8014	8230	8451	8678	8910	9147
640	7949	8163	8381	8604	8833	9067	9306	9551
660	8284	8503	8728	8958	9193	9434	9680	9931
680	8603	8829	9059	9296	9537	9784	10036	10294
700	8930	9162	9399	9641	9889	10142	10400	10665
720	9229	9467	9709	9957	10210	10468	10732	11002
740	9523	9766	10013	10266	10524	10788	11058	11333
760	9834	10082	10336	10594	10858	11128	11403	11684
780	10060	10311	10568	10830	11097	11370	11649	11933
800	10347	10604	10865	11133	11405	11683	11967	12256
820	10588	10849	11114	11385	11662	11944	12231	12524
840	10819	11083	11352	11626	11906	12192	12483	12780
860	11027	11294	11566	11844	12127	12415	12709	13009
880	11220	11490	11766	12046	12332	12623	12920	13222
900	11398	11671	11948	12231	12520	12814	13113	13418
920	11549	11823	12103	12388	12678	12974	13275	13582
940	11686	11962	12244	12531	12823	13120	13423	13732
960	11794	12071	12354	12643	12936	13235	13540	13850
980	11887	12166	12450	12740	13035	13336	13642	13953
1000	11957	12237	12523	12814	13110	13412	13720	14033

TABLE 5
FUGACITY IN BARS

TEMP	PRESSURE IN BARS 8900	9000	9100	9200	9300	9400	9500	9600
20	(8)	(9)	(9)	(10)	(10)	(11)	(12)	(12)
40	18	20	21	22	23	25	26	27
60	37	40	42	44	46	49	52	55
80	69	72	76	80	85	89	94	99
100	117	123	129	136	143	150	157	165
120	187	196	205	215	226	237	248	260
140	284	297	311	325	340	356	372	389
160	410	428	447	467	488	510	532	556
180	570	595	620	647	675	703	733	765
200	763	795	828	862	897	934	973	1013
220	995	1035	1076	1119	1164	1210	1258	1308
240	1261	1310	1361	1413	1468	1525	1584	1645
260	1562	1621	1682	1745	1810	1878	1949	2021
280	1892	1961	2033	2107	2184	2264	2346	2431
300	2253	2333	2416	2502	2591	2683	2778	2876
320	2643	2735	2830	2928	3029	3134	3242	3353
340	3055	3158	3265	3376	3490	3607	3729	3854
360	3482	3598	3717	3840	3966	4097	4231	4370
380	3932	4059	4190	4326	4465	4609	4757	4909
400	4399	4539	4683	4831	4983	5140	5301	5468
420	4872	5024	5180	5340	5505	5675	5850	6029
440	5342	5504	5672	5844	6021	6203	6391	6583
460	5821	5995	6174	6358	6548	6742	6941	7146
480	6291	6476	6666	6861	7062	7268	7479	7696
500	6785	6981	7182	7389	7601	7818	8041	8270
520	7253	7459	7671	7888	8111	8339	8573	8813
540	7684	7899	8120	8346	8578	8815	9059	9308
560	8153	8378	8609	8845	9087	9335	9589	9849
580	8561	8794	9032	9277	9527	9783	10045	10313
600	8971	9211	9458	9710	9968	10232	10502	10779
620	9390	9638	9893	10153	10419	10692	10970	11255
640	9801	10057	10319	10587	10861	11141	11428	11721
660	10189	10452	10721	10996	11277	11564	11858	12158
680	10557	10827	11102	11383	11671	11965	12265	12572
700	10935	11211	11493	11781	12075	12375	12682	12996
720	11278	11559	11847	12141	12440	12747	13059	13378
740	11614	11901	12194	12493	12798	13110	13428	13753
760	11971	12264	12563	12868	13179	13496	13820	14151
780	12223	12519	12822	13130	13444	13765	14092	14426
800	12552	12853	13160	13473	13793	14119	14451	14790
820	12823	13128	13439	13757	14080	14410	14746	15088
840	13082	13391	13705	14026	14353	14686	15025	15371
860	13315	13626	13944	14267	14597	14932	15275	15623
880	13531	13845	14165	14491	14823	15161	15506	15857
900	13728	14045	14367	14695	15030	15370	15717	16070
920	13895	14213	14537	14867	15203	15545	15893	16248
940	14046	14366	14692	15024	15361	15705	16055	16411
960	14165	14487	14814	15147	15486	15831	16182	16540
980	14271	14594	14923	15257	15598	15945	16297	16656
1000	14352	14677	15008	15344	15687	16036	16390	16751

TABLE 5
FUGACITY IN BARS

TEMP	PRESSURE IN BARS 9700	9800	9900	10000
20	(13)	(14)	(15)	(16)
40	29	31	33	34
60	58	61	64	68
80	104	109	115	121
100	174	182	191	201
120	272	285	299	313
140	407	426	445	466
160	580	606	632	660
180	797	831	866	903
200	1054	1098	1142	1189
220	1360	1414	1470	1528
240	1708	1773	1841	1912
260	2097	2175	2256	2339
280	2519	2610	2705	2802
300	2977	3082	3190	3302
320	3469	3588	3710	3837
340	3983	4116	4254	4395
360	4513	4661	4813	4969
380	5066	5228	5395	5566
400	5639	5815	5996	6183
420	6214	6404	6600	6801
440	6781	6984	7193	7407
460	7357	7573	7795	8023
480	7918	8147	8381	8622
500	8505	8747	8994	9248
520	9059	9312	9571	9836
540	9564	9826	10095	10370
560	10115	10388	10667	10954
580	10588	10869	11157	11452
600	11062	11352	11648	11952
620	11547	11845	12150	12462
640	12020	12327	12640	12960
660	12465	12778	13099	13426
680	12885	13205	13532	13866
700	13316	13643	13977	14318
720	13704	14037	14377	14723
740	14084	14422	14767	15119
760	14489	14833	15184	15542
780	14766	15114	15468	15829
800	15136	15488	15847	16214
820	15437	15793	16156	16526
840	15723	16082	16448	16821
860	15978	16340	16708	17084
880	16214	16578	16948	17326
900	16429	16795	17167	17547
920	16609	16976	17350	17730
940	16773	17142	17517	17899
960	16903	17273	17649	18032
980	17021	17393	17771	18155
1000	17119	17492	17872	18258

TABLE 6
FUGACITY COEFFICIENTS

TEMP	PRESSURE IN BARS 100	200	300	400	500	600	700	800
20	0.005	0.003	0.002	0.001	0.001	0.000	0.000	0.000
40	0.006	0.003	0.002	0.001	0.001	0.001	0.000	0.000
60	0.007	0.004	0.002	0.002	0.002	0.001	0.001	0.001
80	0.010	0.005	0.004	0.003	0.002	0.002	0.002	0.002
100	0.016	0.008	0.006	0.004	0.004	0.003	0.003	0.003
120	0.025	0.013	0.009	0.007	0.006	0.005	0.005	0.004
140	0.042	0.022	0.015	0.012	0.010	0.009	0.008	0.007
160	0.067	0.035	0.025	0.019	0.016	0.014	0.013	0.012
180	0.103	0.054	0.038	0.030	0.025	0.022	0.020	0.018
200	0.153	0.081	0.057	0.045	0.038	0.033	0.030	0.027
220	0.220	0.116	0.081	0.064	0.054	0.047	0.042	0.039
240	0.306	0.161	0.113	0.089	0.075	0.065	0.059	0.054
260	0.411	0.216	0.152	0.119	0.100	0.088	0.079	0.072
280	0.537	0.282	0.198	0.156	0.131	0.114	0.103	0.094
300	0.682	0.360	0.252	0.199	0.167	0.146	0.131	0.120
320	0.788	0.445	0.313	0.246	0.206	0.181	0.163	0.150
340	0.817	0.542	0.381	0.300	0.253	0.222	0.199	0.182
360	0.836	0.640	0.453	0.358	0.302	0.264	0.238	0.219
380	0.855	0.703	0.528	0.420	0.355	0.311	0.280	0.257
400	0.873	0.738	0.599	0.481	0.409	0.359	0.325	0.299
420	0.888	0.768	0.653	0.542	0.463	0.409	0.370	0.340
440	0.898	0.794	0.694	0.595	0.517	0.460	0.417	0.384
460	0.911	0.816	0.726	0.641	0.565	0.506	0.462	0.427
480	0.920	0.835	0.753	0.677	0.609	0.551	0.505	0.469
500	0.929	0.851	0.779	0.711	0.647	0.593	0.548	0.512
520	0.937	0.867	0.802	0.739	0.681	0.631	0.588	0.551
540	0.941	0.879	0.819	0.763	0.711	0.665	0.622	0.587
560	0.948	0.891	0.836	0.786	0.739	0.695	0.657	0.623
580	0.951	0.898	0.849	0.804	0.762	0.721	0.684	0.653
600	0.955	0.908	0.864	0.821	0.781	0.744	0.712	0.683
620	0.961	0.917	0.876	0.836	0.803	0.767	0.738	0.708
640	0.965	0.924	0.887	0.854	0.821	0.790	0.760	0.734
660	0.968	0.930	0.895	0.864	0.834	0.806	0.780	0.757
680	0.971	0.937	0.906	0.878	0.849	0.824	0.799	0.777
700	0.974	0.944	0.915	0.888	0.864	0.840	0.816	0.797
720	0.977	0.949	0.923	0.899	0.875	0.853	0.833	0.815
740	0.979	0.954	0.930	0.907	0.887	0.865	0.850	0.830
760	0.983	0.958	0.935	0.915	0.896	0.878	0.862	0.847
780	0.984	0.961	0.938	0.921	0.904	0.885	0.872	0.858
800	0.985	0.966	0.947	0.928	0.913	0.899	0.885	0.872

TABLE 6
FUGACITY COEFFICIENTS

TEMP	PRESSURE IN BARS							
	900	1000	1100	1200	1300	1400	1500	1600
20	0.000	0.000	0.000	0.000	0.000	0.000	0.000	0.000
40	0.000	0.000	0.000	0.000	0.000	0.000	0.000	0.000
60	0.000	0.000	0.000	0.000	0.000	0.000	0.000	0.000
80	0.001	0.001	0.001	0.001	0.001	0.001	0.001	0.001
100	0.002	0.002	0.002	0.002	0.002	0.002	0.002	0.002
120	0.004	0.004	0.004	0.004	0.003	0.003	0.003	0.003
140	0.007	0.007	0.006	0.006	0.006	0.006	0.006	0.005
160	0.011	0.010	0.010	0.010	0.009	0.009	0.009	0.009
180	0.017	0.016	0.015	0.015	0.014	0.014	0.014	0.014
200	0.025	0.024	0.023	0.022	0.021	0.021	0.020	0.020
220	0.036	0.034	0.033	0.031	0.030	0.030	0.029	0.028
240	0.050	0.047	0.045	0.043	0.042	0.041	0.040	0.039
260	0.067	0.064	0.061	0.058	0.056	0.055	0.053	0.052
280	0.088	0.083	0.079	0.076	0.073	0.071	0.069	0.068
300	0.112	0.105	0.100	0.096	0.093	0.090	0.088	0.086
320	0.140	0.131	0.125	0.120	0.115	0.112	0.109	0.107
340	0.170	0.160	0.152	0.146	0.141	0.137	0.133	0.130
360	0.204	0.192	0.182	0.175	0.168	0.163	0.159	0.156
380	0.240	0.226	0.215	0.206	0.199	0.193	0.188	0.184
400	0.279	0.263	0.250	0.240	0.231	0.224	0.219	0.214
420	0.318	0.301	0.286	0.275	0.265	0.257	0.251	0.245
440	0.359	0.340	0.324	0.311	0.300	0.291	0.284	0.278
460	0.400	0.379	0.362	0.348	0.336	0.326	0.318	0.311
480	0.441	0.419	0.400	0.385	0.372	0.362	0.353	0.345
500	0.482	0.459	0.440	0.423	0.410	0.399	0.389	0.381
520	0.521	0.498	0.477	0.461	0.446	0.434	0.424	0.416
540	0.558	0.533	0.512	0.495	0.480	0.468	0.458	0.449
560	0.593	0.570	0.549	0.532	0.517	0.504	0.493	0.484
580	0.625	0.601	0.581	0.563	0.549	0.536	0.525	0.516
600	0.657	0.632	0.612	0.595	0.580	0.567	0.557	0.547
620	0.686	0.662	0.643	0.626	0.611	0.599	0.588	0.579
640	0.711	0.690	0.672	0.656	0.642	0.629	0.619	0.610
660	0.734	0.716	0.698	0.683	0.669	0.657	0.647	0.639
680	0.757	0.739	0.722	0.708	0.695	0.684	0.674	0.666
700	0.777	0.762	0.746	0.733	0.720	0.710	0.700	0.692
720	0.798	0.781	0.767	0.754	0.743	0.733	0.724	0.717
740	0.816	0.800	0.787	0.775	0.764	0.755	0.747	0.740
760	0.830	0.819	0.807	0.796	0.786	0.778	0.770	0.764
780	0.845	0.831	0.820	0.810	0.801	0.793	0.786	0.781
800	0.859	0.848	0.838	0.829	0.821	0.814	0.807	0.802
820		0.861	0.852	0.844	0.837	0.830	0.824	0.820
840		0.874	0.866	0.858	0.852	0.846	0.841	0.837
860		0.885	0.877	0.871	0.865	0.860	0.856	0.852
880		0.895	0.889	0.883	0.878	0.873	0.870	0.867
900		0.905	0.900	0.895	0.890	0.886	0.883	0.880
920		0.913	0.908	0.904	0.900	0.897	0.894	0.892
940		0.921	0.917	0.913	0.910	0.907	0.904	0.903
960		0.928	0.924	0.921	0.918	0.915	0.913	0.912
980		0.935	0.931	0.928	0.926	0.924	0.922	0.921
1000		0.940	0.937	0.935	0.933	0.931	0.930	0.930

TABLE 6
FUGACITY COEFFICIENTS

TEMP	PRESSURE IN BARS 1700	1800	1900	2000	2100	2200	2300	2400
20	0.000	0.000	0.000	0.000	0.000	0.000	0.000	0.000
40	0.000	0.000	0.000	0.000	0.000	0.000	0.000	0.000
60	0.000	0.000	0.000	0.000	0.000	0.000	0.000	0.000
80	0.001	0.001	0.001	0.001	0.001	0.001	0.001	0.001
100	0.002	0.002	0.002	0.002	0.002	0.002	0.002	0.002
120	0.003	0.003	0.003	0.003	0.003	0.003	0.003	0.003
140	0.005	0.005	0.005	0.005	0.005	0.005	0.005	0.005
160	0.009	0.009	0.009	0.009	0.009	0.009	0.009	0.009
180	0.013	0.013	0.013	0.013	0.013	0.013	0.013	0.013
200	0.020	0.020	0.019	0.019	0.019	0.019	0.019	0.019
220	0.028	0.028	0.028	0.027	0.027	0.027	0.027	0.027
240	0.039	0.038	0.038	0.038	0.037	0.037	0.037	0.037
260	0.051	0.051	0.050	0.050	0.050	0.050	0.050	0.050
280	0.067	0.066	0.065	0.065	0.064	0.064	0.064	0.064
300	0.085	0.083	0.083	0.082	0.081	0.081	0.081	0.081
320	0.105	0.104	0.102	0.101	0.101	0.100	0.100	0.100
340	0.128	0.126	0.125	0.123	0.122	0.122	0.121	0.121
360	0.153	0.151	0.149	0.147	0.146	0.145	0.145	0.144
380	0.180	0.178	0.175	0.174	0.172	0.171	0.170	0.170
400	0.210	0.207	0.204	0.202	0.200	0.199	0.198	0.197
420	0.241	0.237	0.234	0.232	0.230	0.228	0.227	0.226
440	0.273	0.268	0.265	0.262	0.260	0.258	0.257	0.256
460	0.306	0.301	0.297	0.294	0.292	0.290	0.288	0.287
480	0.339	0.334	0.330	0.326	0.324	0.322	0.320	0.319
500	0.375	0.369	0.364	0.361	0.358	0.355	0.353	0.352
520	0.409	0.403	0.398	0.394	0.391	0.388	0.386	0.385
540	0.442	0.436	0.431	0.426	0.422	0.420	0.418	0.416
560	0.477	0.470	0.465	0.461	0.456	0.453	0.451	0.450
580	0.508	0.502	0.496	0.492	0.487	0.484	0.482	0.480
600	0.540	0.533	0.528	0.523	0.518	0.515	0.513	0.511
620	0.571	0.565	0.559	0.555	0.549	0.546	0.544	0.542
640	0.602	0.596	0.590	0.586	0.580	0.577	0.575	0.573
660	0.631	0.625	0.620	0.615	0.609	0.606	0.604	0.603
680	0.658	0.652	0.647	0.643	0.637	0.635	0.633	0.631
700	0.686	0.680	0.675	0.671	0.666	0.663	0.661	0.660
720	0.710	0.705	0.700	0.697	0.692	0.689	0.688	0.687
740	0.734	0.729	0.725	0.721	0.717	0.715	0.714	0.713
760	0.758	0.753	0.750	0.747	0.743	0.742	0.740	0.740
780	0.775	0.771	0.768	0.765	0.763	0.761	0.761	0.760
800	0.797	0.794	0.791	0.788	0.787	0.786	0.785	0.785
820	0.816	0.812	0.810	0.808	0.807	0.806	0.806	0.806
840	0.833	0.830	0.828	0.826	0.826	0.826	0.826	0.826
860	0.849	0.847	0.845	0.844	0.844	0.843	0.844	0.845
880	0.864	0.862	0.861	0.860	0.860	0.860	0.861	0.862
900	0.878	0.876	0.875	0.875	0.875	0.876	0.877	0.878
920	0.890	0.889	0.888	0.888	0.888	0.889	0.891	0.892
940	0.901	0.901	0.900	0.901	0.901	0.903	0.904	0.906
960	0.911	0.911	0.911	0.912	0.913	0.914	0.916	0.918
980	0.921	0.921	0.921	0.922	0.924	0.925	0.928	0.930
1000	0.930	0.930	0.931	0.932	0.934	0.936	0.939	0.941

TABLE 6
FUGACITY COEFFICIENTS

	PRESSURE IN BARS							
TEMP	2500	2600	2700	2800	2900	3000	3100	3200
20	0.000	0.000	0.000	0.000	0.000	0.000	0.000	0.000
40	0.000	0.000	0.000	0.000	0.000	0.000	0.000	0.000
60	0.000	0.000	0.000	0.000	0.000	0.000	0.000	0.000
80	0.001	0.001	0.001	0.001	0.001	0.001	0.001	0.001
100	0.002	0.002	0.002	0.002	0.002	0.002	0.002	0.002
120	0.003	0.003	0.003	0.003	0.003	0.004	0.004	0.004
140	0.005	0.006	0.006	0.006	0.006	0.006	0.006	0.006
160	0.009	0.009	0.009	0.009	0.009	0.009	0.009	0.010
180	0.013	0.013	0.014	0.014	0.014	0.014	0.014	0.015
200	0.020	0.020	0.020	0.020	0.020	0.020	0.021	0.021
220	0.028	0.028	0.028	0.028	0.028	0.029	0.029	0.029
240	0.038	0.038	0.038	0.038	0.039	0.039	0.039	0.040
260	0.050	0.050	0.050	0.051	0.051	0.051	0.052	0.052
280	0.064	0.064	0.065	0.065	0.065	0.066	0.066	0.067
300	0.081	0.081	0.081	0.082	0.082	0.083	0.083	0.084
320	0.100	0.100	0.100	0.101	0.101	0.102	0.103	0.104
340	0.121	0.121	0.122	0.122	0.123	0.123	0.124	0.125
360	0.144	0.144	0.145	0.145	0.146	0.146	0.147	0.148
380	0.170	0.170	0.170	0.170	0.171	0.172	0.173	0.174
400	0.197	0.197	0.197	0.198	0.198	0.199	0.200	0.201
420	0.226	0.226	0.226	0.226	0.227	0.228	0.229	0.230
440	0.256	0.255	0.255	0.256	0.256	0.257	0.258	0.260
460	0.287	0.286	0.286	0.287	0.287	0.288	0.289	0.291
480	0.318	0.318	0.318	0.318	0.319	0.319	0.321	0.322
500	0.351	0.351	0.351	0.351	0.352	0.353	0.354	0.355
520	0.384	0.383	0.383	0.384	0.384	0.385	0.387	0.388
540	0.415	0.414	0.414	0.415	0.415	0.416	0.418	0.419
560	0.449	0.448	0.448	0.448	0.449	0.450	0.451	0.453
580	0.479	0.479	0.478	0.479	0.480	0.481	0.482	0.484
600	0.510	0.509	0.509	0.510	0.511	0.512	0.513	0.515
620	0.541	0.541	0.541	0.541	0.542	0.543	0.545	0.547
640	0.572	0.572	0.572	0.573	0.574	0.575	0.577	0.579
660	0.602	0.602	0.602	0.603	0.604	0.605	0.607	0.609
680	0.631	0.631	0.631	0.632	0.633	0.634	0.636	0.639
700	0.660	0.660	0.660	0.661	0.662	0.664	0.666	0.669
720	0.687	0.687	0.687	0.688	0.690	0.692	0.694	0.696
740	0.713	0.713	0.714	0.715	0.717	0.719	0.721	0.724
760	0.740	0.741	0.741	0.743	0.745	0.747	0.750	0.752
780	0.761	0.761	0.762	0.764	0.766	0.768	0.771	0.774
800	0.786	0.787	0.788	0.790	0.792	0.795	0.798	0.801
820	0.807	0.808	0.810	0.812	0.814	0.817	0.820	0.824
840	0.827	0.829	0.831	0.833	0.836	0.839	0.842	0.846
860	0.846	0.848	0.850	0.852	0.855	0.858	0.862	0.866
880	0.864	0.865	0.868	0.871	0.874	0.877	0.881	0.885
900	0.880	0.882	0.885	0.888	0.891	0.895	0.899	0.903
920	0.895	0.897	0.900	0.903	0.907	0.910	0.915	0.919
940	0.909	0.911	0.914	0.918	0.922	0.926	0.930	0.934
960	0.921	0.924	0.927	0.931	0.935	0.939	0.943	0.948
980	0.933	0.936	0.940	0.944	0.948	0.952	0.957	0.961
1000	0.945	0.948	0.952	0.956	0.960	0.964	0.969	0.974

TABLE 6
FUGACITY COEFFICIENTS

TEMP	PRESSURE IN BARS							
	3300	3400	3500	3600	3700	3800	3900	4000
20	0.000	0.000	0.000	0.000	0.000	0.000	0.000	0.000
40	0.000	0.000	0.000	0.000	0.000	0.000	0.000	0.000
60	0.000	0.000	0.000	0.000	0.000	0.000	0.000	0.000
80	0.001	0.001	0.001	0.001	0.001	0.001	0.001	0.001
100	0.002	0.002	0.002	0.002	0.002	0.002	0.002	0.003
120	0.004	0.004	0.004	0.004	0.004	0.004	0.004	0.004
140	0.006	0.006	0.006	0.007	0.007	0.007	0.007	0.007
160	0.010	0.010	0.010	0.010	0.011	0.011	0.011	0.011
180	0.015	0.015	0.015	0.016	0.016	0.016	0.016	0.017
200	0.021	0.022	0.022	0.022	0.023	0.023	0.024	0.024
220	0.030	0.030	0.031	0.031	0.032	0.032	0.033	0.033
240	0.040	0.041	0.041	0.042	0.043	0.043	0.044	0.045
260	0.053	0.054	0.054	0.055	0.056	0.057	0.057	0.058
280	0.068	0.069	0.069	0.070	0.071	0.072	0.073	0.074
300	0.085	0.086	0.087	0.088	0.089	0.090	0.091	0.092
320	0.104	0.105	0.106	0.108	0.109	0.110	0.112	0.113
340	0.126	0.127	0.128	0.130	0.131	0.133	0.134	0.136
360	0.149	0.151	0.152	0.153	0.155	0.157	0.159	0.160
380	0.175	0.176	0.178	0.179	0.181	0.183	0.185	0.187
400	0.203	0.204	0.206	0.207	0.209	0.211	0.213	0.216
420	0.231	0.233	0.235	0.237	0.239	0.241	0.243	0.246
440	0.261	0.263	0.265	0.267	0.269	0.271	0.274	0.276
460	0.292	0.294	0.296	0.298	0.301	0.303	0.306	0.309
480	0.324	0.326	0.328	0.330	0.332	0.335	0.338	0.341
500	0.357	0.359	0.361	0.364	0.366	0.369	0.372	0.376
520	0.390	0.392	0.394	0.397	0.400	0.403	0.406	0.409
540	0.421	0.423	0.426	0.429	0.431	0.435	0.438	0.442
560	0.455	0.457	0.460	0.463	0.466	0.469	0.473	0.477
580	0.486	0.488	0.491	0.494	0.497	0.501	0.504	0.508
600	0.517	0.520	0.523	0.526	0.529	0.533	0.536	0.540
620	0.549	0.552	0.555	0.558	0.562	0.565	0.569	0.573
640	0.581	0.584	0.587	0.590	0.594	0.598	0.602	0.606
660	0.612	0.615	0.618	0.621	0.625	0.629	0.633	0.638
680	0.641	0.644	0.648	0.651	0.655	0.659	0.664	0.668
700	0.671	0.675	0.678	0.682	0.686	0.690	0.695	0.700
720	0.699	0.703	0.706	0.710	0.714	0.719	0.724	0.729
740	0.727	0.730	0.734	0.738	0.743	0.747	0.752	0.757
760	0.756	0.759	0.763	0.767	0.772	0.777	0.782	0.787
780	0.778	0.782	0.786	0.790	0.795	0.800	0.805	0.810
800	0.805	0.808	0.813	0.817	0.822	0.827	0.833	0.838
820	0.828	0.832	0.836	0.841	0.846	0.851	0.857	0.863
840	0.850	0.854	0.859	0.864	0.869	0.874	0.880	0.886
860	0.870	0.875	0.879	0.884	0.890	0.895	0.901	0.907
880	0.889	0.894	0.899	0.904	0.910	0.915	0.921	0.928
900	0.907	0.912	0.917	0.923	0.928	0.934	0.940	0.947
920	0.924	0.929	0.934	0.939	0.945	0.951	0.957	0.964
940	0.939	0.944	0.950	0.955	0.961	0.967	0.974	0.980
960	0.953	0.958	0.964	0.969	0.975	0.981	0.988	0.994
980	0.966	0.972	0.977	0.983	0.989	0.995	1.001	1.008
1000	0.979	0.984	0.990	0.995	1.001	1.007	1.013	1.020

TABLE 6
FUGACITY COEFFICIENTS

	PRESSURE IN BARS							
TEMP	4100	4200	4300	4400	4500	4600	4700	4800
20	0.000	0.000	0.000	0.000	0.000	0.000	0.000	0.000
40	0.000	0.000	0.000	0.000	0.000	0.000	0.000	0.000
60	0.000	0.000	0.000	0.000	0.000	0.000	0.000	0.000
80	0.001	0.001	0.001	0.002	0.002	0.002	0.002	0.002
100	0.003	0.003	0.003	0.003	0.003	0.003	0.003	0.003
120	0.004	0.005	0.005	0.005	0.005	0.005	0.005	0.005
140	0.007	0.008	0.008	0.008	0.008	0.008	0.009	0.009
160	0.011	0.012	0.012	0.012	0.013	0.013	0.013	0.014
180	0.017	0.017	0.018	0.018	0.019	0.019	0.020	0.020
200	0.024	0.025	0.025	0.026	0.027	0.027	0.028	0.028
220	0.034	0.035	0.035	0.036	0.037	0.037	0.038	0.039
240	0.045	0.046	0.047	0.048	0.049	0.050	0.051	0.052
260	0.059	0.060	0.061	0.062	0.064	0.065	0.066	0.067
280	0.075	0.077	0.078	0.079	0.081	0.082	0.083	0.085
300	0.094	0.095	0.097	0.098	0.100	0.102	0.103	0.105
320	0.115	0.116	0.118	0.120	0.122	0.124	0.126	0.128
340	0.138	0.139	0.141	0.143	0.146	0.148	0.150	0.152
360	0.162	0.164	0.167	0.169	0.171	0.174	0.176	0.179
380	0.189	0.192	0.194	0.197	0.199	0.202	0.205	0.208
400	0.218	0.221	0.223	0.226	0.229	0.232	0.235	0.239
420	0.248	0.251	0.254	0.257	0.260	0.264	0.267	0.271
440	0.279	0.282	0.285	0.289	0.292	0.296	0.299	0.303
460	0.312	0.315	0.318	0.322	0.325	0.329	0.333	0.337
480	0.344	0.348	0.351	0.355	0.359	0.363	0.367	0.372
500	0.379	0.383	0.387	0.390	0.395	0.399	0.404	0.408
520	0.413	0.417	0.421	0.425	0.430	0.434	0.439	0.444
540	0.445	0.449	0.454	0.458	0.463	0.467	0.472	0.478
560	0.481	0.485	0.489	0.494	0.499	0.504	0.509	0.514
580	0.512	0.517	0.521	0.526	0.531	0.536	0.542	0.548
600	0.545	0.549	0.554	0.559	0.564	0.570	0.575	0.581
620	0.578	0.583	0.588	0.593	0.598	0.604	0.610	0.616
640	0.611	0.616	0.621	0.626	0.632	0.638	0.644	0.650
660	0.643	0.648	0.653	0.659	0.664	0.670	0.677	0.683
680	0.673	0.679	0.684	0.690	0.696	0.702	0.708	0.715
700	0.705	0.710	0.716	0.722	0.728	0.734	0.741	0.748
720	0.734	0.740	0.745	0.751	0.758	0.764	0.771	0.778
740	0.763	0.768	0.774	0.781	0.787	0.794	0.801	0.808
760	0.793	0.799	0.805	0.811	0.818	0.825	0.832	0.839
780	0.816	0.822	0.828	0.835	0.842	0.849	0.856	0.864
800	0.844	0.850	0.857	0.864	0.871	0.878	0.885	0.893
820	0.869	0.875	0.882	0.888	0.896	0.903	0.911	0.918
840	0.892	0.899	0.905	0.912	0.920	0.927	0.935	0.943
860	0.914	0.920	0.927	0.934	0.942	0.949	0.957	0.965
880	0.934	0.941	0.948	0.955	0.963	0.970	0.978	0.987
900	0.953	0.960	0.967	0.975	0.982	0.990	0.998	1.006
920	0.971	0.978	0.985	0.992	1.000	1.008	1.016	1.024
940	0.987	0.994	1.001	1.009	1.016	1.024	1.032	1.041
960	1.001	1.008	1.015	1.023	1.030	1.038	1.046	1.055
980	1.015	1.022	1.029	1.036	1.044	1.051	1.060	1.068
1000	1.027	1.033	1.041	1.048	1.055	1.063	1.071	1.079

TABLE 6
FUGACITY COEFFICIENTS

TEMP	PRESSURE IN BARS 4900	5000	5100	5200	5300	5400	5500	5600
20	0.000	0.000	0.000	0.000	0.000	0.000	0.000	0.000
40	0.000	0.000	0.000	0.000	0.000	0.000	0.000	0.000
60	0.000	0.000	0.000	0.001	0.001	0.001	0.001	0.001
80	0.002	0.002	0.002	0.002	0.002	0.002	0.002	0.002
100	0.003	0.003	0.003	0.004	0.004	0.004	0.004	0.004
120	0.006	0.006	0.006	0.006	0.006	0.006	0.007	0.007
140	0.009	0.009	0.010	0.010	0.010	0.010	0.011	0.011
160	0.014	0.014	0.015	0.015	0.015	0.016	0.016	0.017
180	0.021	0.021	0.022	0.022	0.023	0.023	0.024	0.025
200	0.029	0.030	0.030	0.031	0.032	0.033	0.034	0.034
220	0.040	0.041	0.042	0.043	0.044	0.045	0.046	0.047
240	0.053	0.054	0.055	0.057	0.058	0.059	0.060	0.062
260	0.069	0.070	0.071	0.073	0.074	0.076	0.078	0.079
280	0.087	0.088	0.090	0.092	0.093	0.095	0.097	0.099
300	0.107	0.109	0.111	0.113	0.115	0.117	0.120	0.122
320	0.130	0.132	0.134	0.137	0.139	0.142	0.144	0.147
340	0.155	0.157	0.160	0.163	0.166	0.168	0.171	0.175
360	0.182	0.185	0.188	0.191	0.194	0.197	0.200	0.204
380	0.211	0.214	0.217	0.221	0.224	0.228	0.232	0.235
400	0.242	0.245	0.249	0.253	0.257	0.261	0.265	0.269
420	0.274	0.278	0.282	0.286	0.290	0.295	0.299	0.304
440	0.307	0.311	0.316	0.320	0.325	0.329	0.334	0.339
460	0.342	0.346	0.351	0.356	0.360	0.366	0.371	0.376
480	0.376	0.381	0.386	0.391	0.396	0.402	0.407	0.413
500	0.413	0.418	0.423	0.429	0.434	0.440	0.446	0.452
520	0.449	0.454	0.460	0.465	0.471	0.477	0.484	0.490
540	0.483	0.489	0.494	0.500	0.506	0.513	0.519	0.526
560	0.520	0.526	0.532	0.538	0.545	0.551	0.558	0.565
580	0.553	0.559	0.566	0.572	0.579	0.586	0.593	0.600
600	0.587	0.594	0.600	0.607	0.614	0.621	0.628	0.636
620	0.622	0.629	0.635	0.642	0.649	0.657	0.664	0.672
640	0.657	0.663	0.670	0.678	0.685	0.692	0.700	0.708
660	0.690	0.697	0.704	0.711	0.719	0.727	0.735	0.743
680	0.722	0.729	0.736	0.744	0.752	0.760	0.768	0.777
700	0.755	0.762	0.770	0.777	0.785	0.794	0.802	0.811
720	0.785	0.793	0.801	0.809	0.817	0.825	0.834	0.843
740	0.815	0.823	0.831	0.839	0.848	0.856	0.865	0.874
760	0.847	0.855	0.863	0.871	0.880	0.889	0.898	0.907
780	0.871	0.879	0.887	0.896	0.905	0.914	0.923	0.932
800	0.901	0.909	0.917	0.926	0.935	0.944	0.953	0.963
820	0.926	0.935	0.943	0.952	0.961	0.970	0.980	0.989
840	0.951	0.959	0.968	0.977	0.986	0.996	1.005	1.015
860	0.974	0.982	0.991	1.000	1.009	1.019	1.028	1.038
880	0.995	1.004	1.013	1.022	1.031	1.041	1.050	1.060
900	1.015	1.024	1.033	1.042	1.051	1.061	1.071	1.081
920	1.033	1.041	1.050	1.060	1.069	1.079	1.089	1.099
940	1.049	1.058	1.067	1.076	1.086	1.095	1.105	1.115
960	1.063	1.072	1.081	1.090	1.100	1.109	1.119	1.129
980	1.076	1.085	1.094	1.103	1.112	1.122	1.132	1.142
1000	1.087	1.096	1.104	1.113	1.123	1.132	1.142	1.152

TABLE 6
FUGACITY COEFFICIENTS

TEMP	PRESSURE IN BARS 5700	5800	5900	6000	6100	6200	6300	6400
20	0.000	0.000	0.000	0.000	0.000	0.000	0.000	0.000
40	0.000	0.000	0.000	0.000	0.000	0.000	0.000	0.000
60	0.001	0.001	0.001	0.001	0.001	0.001	0.001	0.002
80	0.002	0.002	0.002	0.003	0.003	0.003	0.003	0.003
100	0.004	0.004	0.004	0.005	0.005	0.005	0.005	0.005
120	0.007	0.007	0.008	0.008	0.008	0.008	0.009	0.009
140	0.011	0.012	0.012	0.012	0.013	0.013	0.014	0.014
160	0.017	0.018	0.018	0.019	0.019	0.020	0.021	0.021
180	0.025	0.026	0.027	0.027	0.028	0.029	0.030	0.031
200	0.035	0.036	0.037	0.038	0.039	0.040	0.041	0.042
220	0.048	0.049	0.050	0.052	0.053	0.054	0.056	0.057
240	0.063	0.065	0.066	0.068	0.069	0.071	0.073	0.074
260	0.081	0.083	0.085	0.087	0.089	0.091	0.093	0.095
280	0.101	0.103	0.106	0.108	0.110	0.113	0.115	0.118
300	0.124	0.127	0.129	0.132	0.135	0.138	0.141	0.144
320	0.150	0.153	0.156	0.159	0.162	0.165	0.169	0.172
340	0.178	0.181	0.184	0.188	0.191	0.195	0.199	0.203
360	0.207	0.211	0.215	0.219	0.223	0.227	0.231	0.236
380	0.239	0.244	0.248	0.252	0.257	0.261	0.266	0.271
400	0.273	0.278	0.283	0.287	0.292	0.297	0.302	0.308
420	0.309	0.314	0.319	0.324	0.329	0.335	0.340	0.346
440	0.344	0.350	0.355	0.361	0.367	0.372	0.378	0.385
460	0.382	0.387	0.393	0.399	0.405	0.412	0.418	0.425
480	0.419	0.425	0.431	0.437	0.444	0.451	0.458	0.465
500	0.458	0.465	0.471	0.478	0.485	0.492	0.499	0.507
520	0.497	0.503	0.510	0.517	0.525	0.532	0.540	0.548
540	0.533	0.540	0.547	0.555	0.562	0.570	0.578	0.586
560	0.572	0.580	0.587	0.595	0.603	0.611	0.619	0.628
580	0.608	0.615	0.623	0.631	0.639	0.648	0.657	0.665
600	0.643	0.651	0.659	0.668	0.676	0.685	0.694	0.703
620	0.680	0.688	0.697	0.705	0.714	0.723	0.732	0.742
640	0.717	0.725	0.734	0.743	0.752	0.761	0.771	0.780
660	0.751	0.760	0.769	0.778	0.788	0.797	0.807	0.817
680	0.785	0.794	0.803	0.813	0.822	0.832	0.842	0.852
700	0.820	0.829	0.838	0.848	0.858	0.868	0.878	0.888
720	0.852	0.861	0.871	0.880	0.890	0.901	0.911	0.922
740	0.883	0.893	0.903	0.912	0.923	0.933	0.944	0.955
760	0.916	0.926	0.936	0.946	0.957	0.967	0.978	0.989
780	0.942	0.952	0.962	0.972	0.982	0.993	1.004	1.015
800	0.973	0.983	0.993	1.004	1.014	1.025	1.036	1.048
820	0.999	1.010	1.020	1.031	1.041	1.053	1.064	1.075
840	1.025	1.035	1.046	1.057	1.068	1.079	1.090	1.102
860	1.049	1.059	1.070	1.080	1.091	1.103	1.114	1.126
880	1.071	1.081	1.092	1.103	1.114	1.125	1.137	1.149
900	1.091	1.102	1.113	1.124	1.135	1.146	1.158	1.170
920	1.109	1.120	1.131	1.142	1.153	1.164	1.176	1.188
940	1.126	1.136	1.147	1.158	1.169	1.181	1.193	1.205
960	1.139	1.150	1.161	1.172	1.183	1.194	1.206	1.218
980	1.152	1.162	1.173	1.184	1.195	1.206	1.218	1.230
1000	1.162	1.172	1.183	1.193	1.204	1.216	1.227	1.239

TABLE 6
FUGACITY COEFFICIENTS

TEMP	6500	6600	6700	6800	6900	7000	7100	7200
				PRESSURE IN BARS				
20	0.000	0.000	0.000	0.000	0.000	0.000	0.000	0.000
40	0.000	0.000	0.000	0.000	0.000	0.000	0.000	0.001
60	0.002	0.002	0.002	0.002	0.002	0.002	0.002	0.002
80	0.003	0.003	0.003	0.003	0.004	0.004	0.004	0.004
100	0.005	0.006	0.006	0.006	0.006	0.007	0.007	0.007
120	0.009	0.009	0.010	0.010	0.010	0.011	0.011	0.012
140	0.014	0.015	0.015	0.016	0.016	0.017	0.018	0.018
160	0.022	0.022	0.023	0.024	0.025	0.025	0.026	0.027
180	0.031	0.032	0.033	0.034	0.035	0.036	0.037	0.038
200	0.044	0.045	0.046	0.047	0.049	0.050	0.051	0.053
220	0.059	0.060	0.062	0.063	0.065	0.067	0.068	0.070
240	0.076	0.078	0.080	0.082	0.084	0.086	0.089	0.091
260	0.097	0.099	0.102	0.104	0.107	0.109	0.112	0.115
280	0.120	0.123	0.126	0.129	0.132	0.135	0.138	0.141
300	0.147	0.150	0.153	0.156	0.160	0.163	0.167	0.171
320	0.176	0.179	0.183	0.187	0.191	0.195	0.199	0.203
340	0.207	0.211	0.215	0.220	0.224	0.229	0.233	0.238
360	0.240	0.245	0.249	0.254	0.259	0.264	0.270	0.275
380	0.276	0.281	0.286	0.291	0.297	0.303	0.308	0.314
400	0.313	0.319	0.325	0.330	0.336	0.343	0.349	0.356
420	0.352	0.358	0.364	0.371	0.377	0.384	0.391	0.398
440	0.391	0.398	0.404	0.411	0.418	0.425	0.433	0.440
460	0.432	0.439	0.446	0.453	0.461	0.468	0.476	0.484
480	0.472	0.479	0.487	0.495	0.503	0.511	0.519	0.528
500	0.515	0.523	0.531	0.539	0.547	0.556	0.565	0.574
520	0.556	0.564	0.573	0.581	0.590	0.599	0.609	0.618
540	0.595	0.603	0.612	0.621	0.630	0.640	0.650	0.659
560	0.637	0.646	0.655	0.665	0.674	0.684	0.694	0.704
580	0.675	0.684	0.693	0.703	0.713	0.723	0.734	0.744
600	0.713	0.722	0.732	0.742	0.752	0.763	0.774	0.784
620	0.752	0.762	0.772	0.782	0.793	0.803	0.814	0.826
640	0.790	0.800	0.811	0.822	0.832	0.844	0.855	0.866
660	0.827	0.838	0.848	0.859	0.870	0.882	0.893	0.905
680	0.863	0.873	0.884	0.895	0.907	0.918	0.930	0.942
700	0.899	0.910	0.921	0.933	0.944	0.956	0.968	0.980
720	0.933	0.944	0.955	0.967	0.979	0.991	1.0C3	1.016
740	0.966	0.977	0.989	1.000	1.013	1.025	1.037	1.050
760	1.001	1.012	1.024	1.036	1.048	1.061	1.074	1.086
780	1.027	1.039	1.050	1.063	1.075	1.088	1.100	1.113
800	1.059	1.071	1.083	1.096	1.108	1.121	1.134	1.147
820	1.087	1.099	1.111	1.124	1.137	1.149	1.163	1.176
840	1.114	1.126	1.138	1.151	1.164	1.177	1.190	1.203
860	1.138	1.150	1.163	1.175	1.188	1.201	1.215	1.228
880	1.161	1.173	1.186	1.198	1.211	1.225	1.238	1.252
900	1.182	1.194	1.207	1.220	1.233	1.246	1.260	1.273
920	1.200	1.213	1.225	1.238	1.251	1.264	1.278	1.292
940	1.217	1.229	1.242	1.255	1.268	1.281	1.295	1.308
960	1.230	1.243	1.255	1.268	1.281	1.294	1.308	1.322
980	1.242	1.254	1.267	1.280	1.293	1.306	1.319	1.333
1000	1.251	1.263	1.276	1.288	1.301	1.315	1.328	1.342

TABLE 6
FUGACITY COEFFICIENTS

PRESSURE IN BARS

TEMP	7300	7400	7500	7600	7700	7800	7900	8000
20	0.000	0.000	0.000	0.000	0.000	0.000	0.000	0.000
40	0.001	0.001	0.001	0.001	0.001	0.001	0.001	0.001
60	0.002	0.002	0.002	0.002	0.003	0.003	0.003	0.003
80	0.004	0.004	0.004	0.005	0.005	0.005	0.005	0.005
100	0.007	0.008	0.008	0.008	0.008	0.009	0.009	0.009
120	0.012	0.012	0.013	0.013	0.014	0.014	0.015	0.015
140	0.019	0.019	0.020	0.021	0.021	0.022	0.023	0.024
160	0.028	0.029	0.030	0.030	0.031	0.032	0.034	0.035
180	0.040	0.041	0.042	0.043	0.045	0.046	0.047	0.049
200	0.054	0.056	0.057	0.059	0.061	0.062	0.064	0.066
220	0.072	0.074	0.076	0.078	0.080	0.082	0.085	0.087
240	0.093	0.096	0.098	0.101	0.103	0.106	0.109	0.112
260	0.117	0.120	0.123	0.126	0.130	0.133	0.136	0.140
280	0.145	0.148	0.151	0.155	0.159	0.163	0.167	0.171
300	0.175	0.179	0.183	0.187	0.191	0.196	0.200	0.205
320	0.208	0.212	0.217	0.222	0.227	0.232	0.237	0.242
340	0.243	0.248	0.254	0.259	0.265	0.270	0.276	0.282
360	0.281	0.286	0.292	0.298	0.304	0.311	0.317	0.324
380	0.320	0.327	0.333	0.340	0.346	0.353	0.360	0.368
400	0.362	0.369	0.376	0.383	0.391	0.398	0.406	0.414
420	0.405	0.413	0.420	0.428	0.436	0.444	0.452	0.461
440	0.448	0.456	0.464	0.473	0.481	0.490	0.499	0.508
460	0.493	0.501	0.510	0.519	0.528	0.537	0.547	0.556
480	0.537	0.546	0.555	0.564	0.574	0.584	0.594	0.604
500	0.583	0.593	0.602	0.612	0.622	0.633	0.643	0.654
520	0.628	0.638	0.648	0.658	0.669	0.680	0.691	0.702
540	0.670	0.680	0.690	0.701	0.712	0.723	0.735	0.747
560	0.715	0.726	0.737	0.748	0.759	0.771	0.783	0.795
580	0.755	0.766	0.778	0.789	0.801	0.813	0.825	0.838
600	0.796	0.807	0.819	0.831	0.843	0.855	0.868	0.881
620	0.837	0.849	0.861	0.873	0.886	0.898	0.911	0.924
640	0.878	0.890	0.903	0.915	0.928	0.941	0.954	0.968
660	0.917	0.930	0.942	0.955	0.968	0.981	0.995	1.009
680	0.955	0.967	0.980	0.993	1.006	1.020	1.034	1.048
700	0.993	1.006	1.019	1.032	1.046	1.060	1.074	1.088
720	1.028	1.041	1.055	1.068	1.082	1.096	1.110	1.125
740	1.063	1.076	1.090	1.103	1.117	1.132	1.146	1.161
760	1.100	1.113	1.127	1.141	1.155	1.169	1.184	1.199
780	1.127	1.140	1.154	1.168	1.182	1.197	1.212	1.227
800	1.161	1.175	1.189	1.203	1.217	1.232	1.247	1.262
820	1.190	1.203	1.218	1.232	1.247	1.261	1.276	1.292
840	1.217	1.231	1.245	1.260	1.275	1.289	1.305	1.320
860	1.242	1.256	1.271	1.285	1.300	1.315	1.330	1.346
880	1.266	1.280	1.294	1.309	1.324	1.339	1.354	1.370
900	1.287	1.301	1.316	1.330	1.345	1.361	1.376	1.391
920	1.306	1.320	1.334	1.349	1.364	1.379	1.394	1.410
940	1.322	1.337	1.351	1.366	1.381	1.396	1.411	1.427
960	1.336	1.350	1.364	1.379	1.394	1.409	1.425	1.440
980	1.347	1.361	1.376	1.390	1.405	1.421	1.436	1.452
1000	1.356	1.370	1.384	1.399	1.414	1.429	1.445	1.460

TABLE 6
FUGACITY COEFFICIENTS

TEMP	PRESSURE IN BARS							
	8100	8200	8300	8400	8500	8600	8700	8800
20	0.000	0.000	0.000	0.000	0.000	0.000	0.000	0.000
40	0.001	0.002	0.002	0.002	0.002	0.002	0.002	0.002
60	0.003	0.003	0.003	0.003	0.004	0.004	0.004	0.004
80	0.006	0.006	0.006	0.006	0.007	0.007	0.007	0.007
100	0.010	0.010	0.011	0.011	0.011	0.012	0.012	0.013
120	0.016	0.016	0.017	0.018	0.018	0.019	0.020	0.020
140	0.024	0.025	0.026	0.027	0.028	0.029	0.030	0.031
160	0.036	0.037	0.038	0.039	0.041	0.042	0.043	0.045
180	0.050	0.052	0.053	0.055	0.057	0.059	0.060	0.062
200	0.068	0.070	0.072	0.074	0.076	0.079	0.081	0.083
220	0.090	0.092	0.095	0.097	0.100	0.103	0.106	0.109
240	0.115	0.118	0.121	0.124	0.127	0.131	0.134	0.138
260	0.143	0.147	0.151	0.154	0.158	0.163	0.167	0.171
280	0.175	0.179	0.183	0.188	0.193	0.197	0.202	0.207
300	0.210	0.215	0.220	0.225	0.230	0.236	0.241	0.247
320	0.248	0.253	0.259	0.265	0.271	0.277	0.284	0.290
340	0.288	0.294	0.301	0.308	0.314	0.321	0.328	0.336
360	0.330	0.337	0.345	0.352	0.359	0.367	0.375	0.383
380	0.375	0.383	0.391	0.399	0.407	0.415	0.424	0.433
400	0.422	0.430	0.439	0.448	0.456	0.466	0.475	0.485
420	0.470	0.479	0.488	0.497	0.507	0.517	0.527	0.537
440	0.517	0.527	0.537	0.547	0.557	0.567	0.578	0.589
460	0.566	0.576	0.587	0.597	0.608	0.619	0.631	0.642
480	0.614	0.625	0.636	0.647	0.659	0.670	0.682	0.694
500	0.665	0.676	0.688	0.700	0.712	0.724	0.737	0.749
520	0.714	0.725	0.737	0.750	0.762	0.775	0.788	0.801
540	0.759	0.771	0.783	0.796	0.809	0.822	0.836	0.849
560	0.807	0.820	0.833	0.846	0.860	0.873	0.887	0.902
580	0.850	0.863	0.877	0.890	0.904	0.918	0.932	0.947
600	0.894	0.907	0.921	0.935	0.949	0.963	0.978	0.993
620	0.938	0.952	0.966	0.980	0.994	1.009	1.024	1.040
640	0.981	0.995	1.010	1.024	1.039	1.054	1.070	1.085
660	1.023	1.037	1.052	1.066	1.082	1.097	1.113	1.129
680	1.062	1.077	1.092	1.107	1.122	1.138	1.154	1.170
700	1.103	1.117	1.132	1.148	1.163	1.179	1.196	1.212
720	1.139	1.155	1.170	1.185	1.201	1.217	1.234	1.250
740	1.176	1.191	1.206	1.222	1.238	1.255	1.271	1.288
760	1.214	1.230	1.245	1.261	1.278	1.294	1.311	1.328
780	1.242	1.258	1.273	1.289	1.306	1.322	1.339	1.356
800	1.277	1.293	1.309	1.325	1.342	1.359	1.376	1.393
820	1.307	1.323	1.339	1.355	1.372	1.389	1.406	1.423
840	1.336	1.352	1.368	1.384	1.401	1.418	1.435	1.452
860	1.361	1.377	1.394	1.410	1.427	1.444	1.461	1.478
880	1.385	1.401	1.418	1.434	1.451	1.468	1.485	1.503
900	1.407	1.423	1.440	1.456	1.473	1.490	1.507	1.525
920	1.426	1.442	1.458	1.475	1.492	1.509	1.526	1.543
940	1.443	1.459	1.475	1.492	1.509	1.526	1.543	1.560
960	1.456	1.472	1.489	1.505	1.522	1.539	1.556	1.574
980	1.468	1.484	1.500	1.517	1.534	1.551	1.568	1.586
1000	1.476	1.492	1.509	1.526	1.542	1.560	1.577	1.595

TABLE 6
FUGACITY COEFFICIENTS

PRESSURE IN BARS

TEMP	8900	9000	9100	9200	9300	9400	9500	9600
20	(0.000)	(0.001)	(0.001)	(0.001)	(0.001)	(0.001)	(0.001)	(0.001)
40	0.002	0.002	0.002	0.002	0.003	0.003	0.003	0.003
60	0.004	0.004	0.005	0.005	0.005	0.005	0.006	0.006
80	0.008	0.008	0.008	0.009	0.009	0.010	0.010	0.010
100	0.013	0.014	0.014	0.015	0.015	0.016	0.017	0.017
120	0.021	0.022	0.023	0.023	0.024	0.025	0.026	0.027
140	0.032	0.033	0.034	0.035	0.037	0.038	0.039	0.041
160	0.046	0.048	0.049	0.051	0.053	0.054	0.056	0.058
180	0.064	0.066	0.068	0.070	0.073	0.075	0.077	0.080
200	0.086	0.088	0.091	0.094	0.097	0.099	0.102	0.106
220	0.112	0.115	0.118	0.122	0.125	0.129	0.133	0.136
240	0.142	0.146	0.150	0.154	0.158	0.162	0.167	0.171
260	0.176	0.180	0.185	0.190	0.195	0.200	0.205	0.211
280	0.213	0.218	0.223	0.229	0.235	0.241	0.247	0.253
300	0.253	0.259	0.266	0.272	0.279	0.285	0.292	0.300
320	0.297	0.304	0.311	0.318	0.326	0.333	0.341	0.349
340	0.343	0.351	0.359	0.367	0.375	0.384	0.393	0.401
360	0.391	0.400	0.409	0.417	0.427	0.436	0.445	0.455
380	0.442	0.451	0.461	0.470	0.480	0.490	0.501	0.511
400	0.494	0.504	0.515	0.525	0.536	0.547	0.558	0.570
420	0.548	0.558	0.569	0.581	0.592	0.604	0.616	0.628
440	0.600	0.612	0.623	0.635	0.648	0.660	0.673	0.686
460	0.654	0.666	0.679	0.691	0.704	0.717	0.731	0.744
480	0.707	0.720	0.733	0.746	0.759	0.773	0.787	0.802
500	0.762	0.776	0.789	0.803	0.817	0.832	0.847	0.862
520	0.815	0.829	0.843	0.857	0.872	0.887	0.902	0.918
540	0.863	0.878	0.892	0.907	0.922	0.938	0.954	0.970
560	0.916	0.931	0.946	0.961	0.977	0.993	1.009	1.026
580	0.962	0.977	0.993	1.008	1.024	1.041	1.057	1.074
600	1.008	1.024	1.039	1.055	1.072	1.089	1.106	1.123
620	1.055	1.071	1.087	1.104	1.120	1.137	1.155	1.172
640	1.101	1.118	1.134	1.151	1.168	1.185	1.203	1.221
660	1.145	1.161	1.178	1.195	1.213	1.230	1.248	1.267
680	1.186	1.203	1.220	1.237	1.255	1.273	1.291	1.310
700	1.229	1.246	1.263	1.281	1.298	1.317	1.335	1.354
720	1.267	1.284	1.302	1.320	1.338	1.356	1.375	1.394
740	1.305	1.322	1.340	1.358	1.376	1.395	1.414	1.433
760	1.345	1.363	1.381	1.399	1.417	1.436	1.455	1.474
780	1.373	1.391	1.409	1.427	1.446	1.464	1.483	1.503
800	1.410	1.428	1.446	1.465	1.483	1.502	1.521	1.541
820	1.441	1.459	1.477	1.495	1.514	1.533	1.552	1.572
840	1.470	1.488	1.506	1.525	1.543	1.562	1.582	1.601
860	1.496	1.514	1.532	1.551	1.570	1.589	1.608	1.627
880	1.520	1.538	1.557	1.575	1.594	1.613	1.632	1.652
900	1.543	1.561	1.579	1.597	1.616	1.635	1.654	1.674
920	1.561	1.579	1.598	1.616	1.635	1.654	1.673	1.693
940	1.578	1.596	1.615	1.633	1.652	1.671	1.690	1.710
960	1.592	1.610	1.628	1.646	1.665	1.684	1.703	1.723
980	1.603	1.622	1.640	1.658	1.677	1.696	1.716	1.735
1000	1.613	1.631	1.649	1.668	1.687	1.706	1.725	1.745

TABLE 6
FUGACITY COEFFICIENTS

TEMP	PRESSURE IN BARS 9700	9800	9900	10000
20	(0.001)	(0.001)	(0.001)	(0.002)
40	0.003	0.003	0.003	0.003
60	0.006	0.006	0.006	0.007
80	0.011	0.011	0.012	0.012
100	0.018	0.019	0.019	0.020
120	0.028	0.029	0.030	0.031
140	0.042	0.043	0.045	0.047
160	0.060	0.062	0.064	0.066
180	0.082	0.085	0.088	0.090
200	0.109	0.112	0.115	0.119
220	0.140	0.144	0.148	0.153
240	0.176	0.181	0.186	0.191
260	0.216	0.222	0.228	0.234
280	0.260	0.266	0.273	0.280
300	0.307	0.315	0.322	0.330
320	0.358	0.366	0.375	0.384
340	0.411	0.420	0.430	0.440
360	0.465	0.476	0.486	0.497
380	0.522	0.533	0.545	0.557
400	0.581	0.593	0.606	0.618
420	0.641	0.653	0.667	0.680
440	0.699	0.713	0.727	0.741
460	0.758	0.773	0.787	0.802
480	0.816	0.831	0.847	0.862
500	0.877	0.893	0.908	0.925
520	0.934	0.950	0.967	0.984
540	0.986	1.003	1.020	1.037
560	1.043	1.060	1.078	1.095
580	1.092	1.109	1.127	1.145
600	1.140	1.158	1.177	1.195
620	1.190	1.209	1.227	1.246
640	1.239	1.258	1.277	1.296
660	1.285	1.304	1.323	1.343
680	1.328	1.347	1.367	1.387
700	1.373	1.392	1.412	1.432
720	1.413	1.432	1.452	1.472
740	1.452	1.472	1.492	1.512
760	1.494	1.514	1.534	1.554
780	1.522	1.542	1.562	1.583
800	1.560	1.580	1.601	1.621
820	1.591	1.612	1.632	1.653
840	1.621	1.641	1.661	1.682
860	1.647	1.667	1.688	1.708
880	1.672	1.692	1.712	1.733
900	1.694	1.714	1.734	1.755
920	1.712	1.732	1.752	1.773
940	1.729	1.749	1.769	1.790
960	1.743	1.763	1.783	1.803
980	1.755	1.775	1.795	1.815
1000	1.765	1.785	1.805	1.826